T0177461

The Biology of Death

The Biology of Death

How Dying Shapes Cells, Organisms, &
Populations

GARY C. HOWARD

OXFORD
UNIVERSITY PRESS

OXFORD
UNIVERSITY PRESS

Oxford University Press is a department of the University of Oxford. It furthers
the University's objective of excellence in research, scholarship, and education
by publishing worldwide. Oxford is a registered trade mark of Oxford University
Press in the UK and certain other countries.

Published in the United States of America by Oxford University Press
198 Madison Avenue, New York, NY 10016, United States of America.

Library of Congress Cataloging-in-Publication Data
Names: C. Howard, Gary, author.
Title: The biology of death : how *dying shapes cells, organisms, & populations* / Gary C. Howard.
Description: New York, NY : Oxford University Press Academic Newgen,
[2021] | Includes bibliographical references and index.
Identifiers: LCCN 2021011285 (print) | LCCN 2021011286 (ebook) |
ISBN 9780190687724 (hardback) | ISBN 9780190687748 (epub)
Subjects: LCSH: Death (Biology)
Classification: LCC QH530 .C46 2021 (print) | LCC QH530 (ebook) |
DDC 571.9/39—dc23
LC record available at https://lccn.loc.gov/2021011285
LC ebook record available at https://lccn.loc.gov/2021011286

DOI: 10.1093/oso/9780190687724.001.0001

1 3 5 7 9 8 6 4 2

Printed by Sheridan Books, Inc., United States of America

For Shirley, Rebecca, and Amanda, and all of our ancestors and descendants

About the Dead Man and Fungi

Marvin Bell

The dead man has changed his mind about moss and mold.

About mildew and yeast.

About rust and smut, about soot and ash.

Whereas once he turned from the sour and the decomposed, now he
breathes deeply in the underbelly
       of the earth.

Of mushrooms, bakers yeast, fungi of wood decay, and the dogs
preceding their masters to the
       burnt acre of morels.

And the little seasonals themselves, stuck on their wobbly pin stems.

For in the pan they float without crisping.

For they are not without a hint of the sublime, nor the curl of a hand.

These are the caps and hairdos, the mini-umbrellas, the zeppelins of a
world in which human
       beings are heavy-footed mammoths.

Puffballs and saucers, recurrent, recumbent, they fill the encyclopedia.

Not wrought for the pressed eternity of flowers or butterflies.

Loners and armies alike appearing overnight at the point of return.

They live fast, they die young, they will be back.

Credit: Marvin Bell, "About the Dead Man and the Fungi" from "The Book of the Dead Man
(Fungi) in Vertigo: The Living Dead Man Poems. Copyright © 2011 by Marvin Bell. Reprinted
with the permission of The Permissions Company, LLC on behalf of Copper Canyon Press,
coppercanyonpress.org.

Contents

Preface

The words *death* and *dying* are uncomfortable. They conjure up fear of the unknown, loss, and sadness. We know on an intellectual level that everyone dies, but we avoid thinking about it. We make jokes about it. We really want nothing to do with it. For the last several decades, death has been almost invisible in the lives of people in developed countries. Those who die are whisked away. "Viewing" of the body is held at the funeral home. It is said that the person has "passed" rather than died. Humans are conflicted by their instinct to avoid death at the same time that they are aware that death is inevitable—an awareness that psychologists refer to as mortality salience.

All living things eventually die. Yet death is much more complicated than simply the last event of life. Death is actually interwoven into life at many levels. It has influenced what we are as a species. In many ways, we depend on death for our very existence. In fact, death has shaped most of the aspects of our life, and it is intimately linked with our growth, development, protection against disease, and more.

In a broader sense, a careful examination of death involves a multitude of fascinating issues: How do we define life and death? Are viruses alive or not? How do we know when a person is dead? Why do we age and can we do anything about it? It also raises a host of ethical questions about research into aging and if it is even ethical to consider extending life spans. Given the pace of medical advances, what is the future of death? In the same way, death has influenced the direction of entire species. By foreclosing some evolutionary pathways, it has shaped all life through a cycle of life and death, both throughout time and in normal development. The loss of populations and even species channeled evolution and eventually focused human evolution to modern humans.

Most amazingly of all, living organisms evolved systems to use the death of cells to their own advantage. All cells seem to carry "death" gene programs that can be activated as needed. Many organisms, including humans, now rely on those programs to kill certain cells during development and to battle disease. The most dramatic occurs during metamorphosis of insects and frogs, but humans use the death of specific cells to hone our immune system and give us fingers and toes. Even single-celled organisms use quorum sensing to eliminate some cells to ensure the overall survival of the colony in harsh environments. Plants use it to cause leaves and fruit to drop in the fall.

Thus, there is more to death than just dying. It is not simply the end of an individual. Death is intimately intertwined with life. Since the beginning, life has had to contend with death, but the most amazing aspect is that life evolved ways to use death. It has made us what we are today by acting at multiple levels, including cellular, tissue, individual, and even population. Most amazingly, living organisms, even many single-celled organisms, have learned how to take advantage of death to promote growth, development, protection from diseases, and even life itself. Death is more than just dying.

This book took several years to think about and prepare, and I didn't do it alone. I am very grateful to a number of people. I thank Debra Bakerjian (University of California, Davis), Henry Gilbert (California State University, East Bay and University of California, Berkeley), Birgit Schilling (Buck Institute for Research on Aging), and Andrey Tsvetkov (University of Texas, Houston) for their insightful and helpful comments on the manuscript. I also thank Jeremy Lewis for his guidance and patience and the extraordinary Oxford University Press team, including Bronwyn Geyer, for their skill and generosity and the outstanding work of Archanaa Rajapandian and her team at NewGen. Finally, I thank my wife and daughters for giving me the time and support to complete this project.

1

Death in Life

Dying is a very dull, dreary affair. And my advice to you is to have
nothing whatever to do with it.

—W. Somerset Maugham

Death is serious business. So why do we laugh about it so much? We make up
funny expressions about it: Box city. Boot Hill. Worm farm. Pushing up daisies.
Kick the bucket. Bite the dust. Other cultures do the same. In German, it might
be "den Loeffel abgeben" (hand over the spoon) or "ins Grass beissen" (bite into
the grass). We whistle past the graveyard. It's easy to make jokes about death. We
laugh to save ourselves from thinking about it. Humor is one of our best defenses
against the thought of death.

We need that defense because death is no laughing matter. It conjures up fear,
uncertainty, and grief. Its finality intimidates us. Our fear of it might be innate.
Death is the most-feared topic among children and adolescents. Death anxiety—
or thanatophobia, as it is formally known—affects children as young as four years
of age, and it is one of the greatest fears of adolescents (Lane and Gullone 1999).
Interestingly, the fear of death decreases for those closest to it. The aged, especially those over sixty-five, have less fear of death than other age groups. We also
fear the death of others. Psychologists use the Holmes and Rahe scale (Holmes
and Rahe 1967) to measure stress. On that scale, the death of a spouse or, perhaps
even worse, the death of a child is rated as by far the most stressful event in life.

Some psychologists refer to this fear as mortality salience (Burke, Martens,
and Faucher 2010). We are conflicted by our instinct to avoid death at the same
time as we know intellectually that death is inevitable. The fear of death fits into
the larger theory of terror management, which assumes that almost all of our
activities are driven by a fear of dying. We defend ourselves by avoiding death or
distracting themselves from it, perhaps by ignoring it or laughing about it.

Death isn't just scary. It raises questions about the meaning of life and our
ultimate place in the universe. It also causes us to realize that we are still biological beings. It's easy to forget our connections to the natural world. Most
of us, at least in much of the developed world, live in a protective bubble.
We are mostly protected from the elements. We are warmed in the winter

The Biology of Death. Gary C. Howard, Oxford University Press. © Oxford University Press 2021.
DOI: 10.1093/oso/9780190687724.003.0001

Figure 1.1 Expulsion from paradise. Hans Holbein (1497–1543) created a series of woodblock prints that depicted various scenes in which Death intrudes in everyday life. In this scene, Adam and Eve are cast out of Eden to work "till thou return unto the ground; for out of it wast thou taken: for dust thou *art*, and unto dust shalt thou return" (Genesis 3:19).

and cooled in the summer. We no longer need to fear predators. We don't have to forage for food; it's readily available at a local supermarket. We go to work, watch television, text with each other, and often forget the natural world that still exists all around us. Death, like birth and sickness, is one of those fundamental events that force us to remember that we are still biological animals.

It's a cliché to say that death is a part of life. The Austrian British philosopher Ludwig Wittgenstein said, "Death is not an event of life. Death is not lived through" (1922, 88). We mostly think of death as the event that ends life. Everyone dies, and so we naturally associate death with the end of the life of a person, or a beloved pet, or the flowers at the end of summer. It could be our own life or that of a loved one. In all of these cases, death is seen as an end. We mourn the lost person or pet. We miss the beauty of the flowers as they wilt, the color fades, and the blooms fall.

Yet death is much more complicated than simply the last event of life. Death is actually interwoven into life at many levels. Some of its manifestations represent an end. Others are critical for development and health.

Rather than having nothing to do with it, as suggested by Maugham's quote at the beginning of this chapter, we depend on death for our very existence. In fact, death has shaped most of the aspects of our life, and it is intimately linked with our growth, development, protection against disease, and more.

In this book, we will examine the biology of death from several perspectives. We will begin by looking at what life and death are. How do we define life and death? Which organisms are living and which are not? For example, viruses seem to exist at the boundary of life and death (see Chapter 2). In particular, the definition of death has profound practical implications. The traditional definition is the loss of activity in the heart. However, modern medicine can work miracles to artificially maintain a heartbeat and breathing. Many physicians and ethicists prefer brain death as a standard, but that brings many complications, especially legal and ethical ones.

We will then examine the death of whole organisms and the events that occur before and after death (Chapters 3 and 4). So why do we age and die? Deaths from trauma or disease are clear. But what is "old age"—and why do we age at all? We might speculate that old people die off to spare scarce resources for the benefit of a younger generation. That explanation might make us feel better about aging and death; it might seem to provide some purpose to these events. But as comforting as that explanation might be, it is simply wrong. Aging and death generally occur well after the reproductive years, and thus there is no way for evolutionary pressure to select for this outcome.

We all die, but what happens in that process (Chapter 5). The result is invariably "ashes to ashes and dust to dust." Death in plants and animals sets in motion a well-defined series of biochemical and microbiological events, and those events are so predictable that they are frequently used in forensic investigations (Chapter 6). When the temperature and other factors are taken into consideration, the lengths of time after death when rigor mortis sets in and then relaxes are well known. Even the succession of insects that invade the dead body is known, and those insects too die and decompose in a predictable manner.

For many thousands of years, humans have recognized the importance of death. They treated the dead in special ways. They buried dead people in locations as simple as a cave or as complex as a pyramid; valuable or useful objects were often included in the burials. Humans are the only species known to show such caring for the dead. Humans have been doing that for a long time, and many anthropologists believe that caring for the dead is a defining characteristic of being human.

The events leading up to death are also well studied. In some cases, the cause of death is trauma or illness. The courses of diseases are, in many cases, understood and reasonably predictable. Aging is another story. Some of its features are trivial (e.g., gray hair, loss of hair, wrinkles). Others are more problematic and involve loss of vigor and strength, muscle and bone mass, hearing and vision, and mental capacity. Americans and many others spend amazing amounts of money trying to slow these inevitable effects.

But why do we age at all? Are we some kind of biological machine that wears out over time? The short answer is that no one really knows. Genetics is clearly involved (Chapter 7). Longevity and certain disease conditions that limit longevity run in families. Progeria is a rare condition that tragically accelerates aging, so that even young children exhibit accelerated manifestations of aging. Might there be a real "fountain of youth" that would allow us to live longer and healthier, even forever? Are red wine and chocolate good for us as well as just good? Molecular genetics has made remarkable progress in the last decades. What was once science fiction is now becoming science fact.

Moreover, the events immediately before and after the death of an individual are not the whole story. Death affects many other aspects of our life and development. Death—the death of cells, tissues, individuals, and even whole populations—is actually interwoven into the lives of all living things. It has made us what we are. The deaths of cells, individuals, and populations have formed us. In fact, death has shaped all life through cycles of life and death, through evolution, and through normal and pathological development. Recycled organic material from one generation provides the nutrients for the next. The atoms and molecules of which we are made have been used before by other living organisms. We "borrow" them for a brief period and, eventually, must give them back so they can be used again and again.

Amazingly, organisms have evolved systems that use death to their advantage. Every cell carries the genetic program that would ensure its death (Chapters 8-11). In fact, normal development and life could not exist without the carefully regulated death of certain cells. At the cellular and tissue levels, normal development relies on the death of some cells to ensure the proper organization of others. For example, caterpillars transform into butterflies during metamorphosis; during this process, cells die and tissues are lost even as new tissues are built from new cells. Some aspects of human development bear some resemblance to metamorphosis; for example, a fetus in the early stages of development has tissue between the fingers, but as the fetus grows, that tissue dies so that the digits can move independently. Another example of the role of this process of programmed cell death, or apoptosis, can be seen when the human body is afflicted by disease; damaged cells die off, which allows other healthy cells and the whole organism to survive. A third example of how cell death is part of human life is the fact that certain types of cells wear out and die and must be replaced regularly. In plants, a similar process is a protection that limits infections. Also, leaves fall from trees in the autumn because a line of cells die where a leaf stem is attached.

Even evolution has been influenced by death (Chapter 12). Darwin discovered the principle of natural selection or "survival of the fittest"; in essence, those who are more fit reproduce more effectively. The less fit reproduce less effectively and die sooner. In Farley Mowat's book *Never Cry Wolf* (2010), the main character learns that the wolves in the Canadian Artic act as a tool for nature to cull the caribou herd. The wolves observe the herd and test its members to determine which ones are weak, sick, old, or otherwise less able to defend themselves. Those are the caribou that they attack, and in doing that, they eliminate those weaker members, leaving the stronger ones to survive and reproduce. Without the pressure of the wolves, the caribou herd would become weaker overall and suffer. By eliminating injured, old, and diseased animals, the wolves ensure the survival of the caribou. Darwin's mechanism of survival of the fittest or natural selection has been translated into action. The weaker animals are killed and the more fit survive to procreate. Death ensures life.

In the same way, death has influenced the direction of entire species by foreclosing some evolutionary pathways. Many species have died out over the course of time. And at several points in the history of life on earth nearly every living thing was killed off by cataclysmic events: even highly successful species were suddenly killed off, leaving only a small number to carry on. Whole dominant groups were wiped out. For example, the dinosaurs had ruled earth for millions of years. In a sudden (at least in terms of geologic time) event, they were killed off, and mammals began their rise. So too many hominoid species were lost over time by death or interbreeding, leaving only modern humans (Chapter 13).

The study of death raises significant issues of bioethics (Chapter 14). Research might be directly related to extending human life, or it may focus simply on treating or preventing diseases, particularly of the aged. In either case, thorny ethical problems involve fairness, cost, the risk of unintended consequences, and the alleviation of suffering. Although this book does not examine the many important religious or philosophical aspects of death, these ethical issues are critical for scientists and the general public to consider.

Finally, we will consider the future of death (Chapter 15). In a world in which many of us in the West are insulated from many of the consequences of our biological nature, several events remind us that we are still living organisms: birth, hunger, injury, sickness, and aging. Death is final, at least for now. However, exciting new advances in molecular genetics, stem cell biology, and regenerative medicine may let us stave off death for a while.

Thus, there is more to death than just dying (Chapter 16). It is not simply the end of an individual. Death is intimately intertwined with life. It has made us what we are today by acting at multiple levels: cells, tissues, individuals, even entire populations. Most amazingly, living organisms, even many single-cell organisms, have learned how to take advantage of death to promote growth, development, protection from disease, and even life itself.

2

Defining Life and Death

> The boundaries which divide Life from Death are at best shadowy
> and vague. Who shall say where the one ends, and where the other
> begins?
>
> —Edgar Allan Poe, "The Premature Burial"

In his famous short story "The Premature Burial," Edgar Allan Poe vividly describes the horror of being buried alive. The premise was that some people are declared dead even though they are still alive, and after burial they awaken in the grave. To give his story credence, Poe cited several accounts (including those of a congressman's wife, a Parisian journalist, and an artillery officer) of such mistakes, in which graves were opened to reveal that the persons buried had struggled to get out of the grave. He wrote, "To be buried while alive is, beyond question, the most terrific of these extremes which has ever fallen to the lot of mere mortality. That it has frequently, very frequently, so fallen will scarcely be denied by those who think."

The story and the accounts cited were fiction, but the fear of premature burial was widespread in the eighteenth and nineteenth centuries, and those fears were supported by numerous anecdotal accounts that were likely no more accurate than the ones Poe used. "I might refer at once, if necessary, to a hundred well authenticated instances," the narrator of Poe's story says. Of course, the "authenticated" instances were completely fictional.

While the stories were not true and might sound silly to us now, they were common enough in the day: somebody knew somebody who told them that they'd seen it. Some people even made provisions to help them escape the grave or to signal that they were still alive to those above. For example, J. G. Krichbaum of Ohio received a US patent for a device that would alert others to help those who had been buried alive: "My invention has relation to that class of devices for indicating life in persons buried under doubt of being in a trance in which connection is obtained between the person in the grave and an indicator on the ground over the grave through a casing or tube leading from the surface of the ground down to the coffin" (Krichbaum 1882). His and other devices were thought to be needed because of mistakes by those responsible for determining

The Biology of Death. Gary C. Howard, Oxford University Press. © Oxford University Press 2021.
DOI: 10.1093/oso/9780190687724.003.0002

if a person was truly dead or not. The fear was that the "deceased" might be in some sort of coma and just appear to be dead. The idea even gained traction with the scientific establishment. The London Association for the Prevention of Premature Burial was established in 1896 by Walter Hadwen and William Tebb (*JAMA* 1899), and Tebb and Colonel Edward Perry Vollum published *Premature Burial and How It May Be Prevented* (Tebb and Vollum 1896).

To Poe and others of that era, "the boundaries which divide Life and Death are at best shadowy and vague." Of course, medical science in Poe's time was much less sophisticated than it is today. The processes of death were then poorly understood, and instrumentation for proper testing was nonexistent. Death was determined by the standard of that day—the loss of breathing and a heartbeat. Interestingly, even with all of our modern medical knowledge and instrumentation, that is still the most-used method of determining death today.

Medicine has come a long way since Poe's time, but our increased knowledge of life and death and the sophisticated instrumentation that we have developed have not completely ended the uncertainty of determining whether someone has really died, at least in some cases. Questions about life and death continue to haunt us. Premature burial is not the problem; now we wonder how we can be sure that a person is really dead before we harvest organs for transplantation or whether it is too soon to give up and terminate life support for a patient in a persistent vegetative state. Families hoping against hope for a miracle sometimes resist a medical determination. Who can blame them? Furthermore, our powerful modern methods for artificially maintaining respiration and circulation have clouded the determination of death. New tests, such as those for brain activity, are widely used and just as widely disputed. More about this later. But the issue in Poe's time and, in fact, still in our own is how to determine who is actually dead. To resolve that, we need to understand what it means to be dead or alive.

It seems like it should be easy to differentiate between things that are alive and things that are not, and in many cases it is. We are unlikely to make a mistake when we say that a playful kitten is alive or that a pebble is not. Yet we use the terms *life* and *death* so casually that we tend to take them for granted. They are so familiar that we can easily lose track of what they really mean. Most of the time it doesn't matter. But sometimes it does.

The terms *living* and *dead* refer to very different things, of course. Let's start with the second one. Nonliving things can be divided into two categories: things that were never alive and things that were once alive but no longer are. It's the latter group of things that we call *dead*. Things that were never alive and things that are dead share several characteristics, including a lack of movement and a lack of the abilities to reproduce and use energy. Some things that were once alive, such as fossils, retain a degree of internal organization, but so do crystals, which were never alive.

But when it comes to things that are living and things that are dead, sometimes the differentiation is less clear. Are leaves that have fallen from trees in autumn dead or still alive?

Life: Hard to Define

In ancient Greece, Aristotle wrote that "nature proceeds little by little from things lifeless to animal life in such a way that it is impossible to determine the exact line of demarcation" (Aristotle 1910: bk. VIII, pt. 1). We could paraphrase his statement in referring to death. Nature proceeds little by little from things alive to death in such a way that it is impossible to determine the exact line of demarcation.

Since death is the absence of life in something that was once alive, the question becomes: What is life? What is it that makes something alive? What is so special about life that we can differentiate it from death? Scientists have long struggled with these questions. Until the nineteenth century, vitalism was the dominant explanation for life. Vitalism posited that living organisms were governed by nonphysical principles. They were fundamentally different from nonliving material. For example, they argued that organic material could not be made with nonliving materials. Vitalism suffered a serious blow in 1828 when Friedrich Wöhler synthesized urea from inorganic materials. In more recent years, many scientists have weighed in with varying degrees of success, but in the end, all definitions of life seem to come up short.

The physicist Erwin Schrödinger published his classic book *What Is Life?* in 1944. Schrödinger saw life as bringing "order from disorder." Heredity is a key characteristic of life and one aspect of bringing order from disorder. How are traits passed on from one generation to the next? Schrödinger believed that whatever was responsible for heredity had to be extremely small, permanent, and nonrepetitive. At that time, little was known about the biochemical basis of heredity. Classical genetics was established well after earlier work by Mendel and others was rediscovered in the early 1900s. The early geneticists who worked with fruit flies in the genus *Drosophila*, such as Thomas Hunt Morgan, Alfred Sturtevant, H. J. Muller, and Calvin Bridges, made remarkable—even astounding—progress on understanding genetics by sophisticated crossing of flies and examination of offspring (Rubin, 1988). However, without knowing what molecules were actually involved in the process, it was not possible to connect classic genetics to the contents of the cell. Interestingly, also in 1944, Oswald Avery solved one of the biggest problems in biology by showing that DNA was, in fact, the hereditary material. So Schrodinger's book did not really answer its own question. Yet that does not mean that the book lacked value. The real value of Schrodinger's ideas about life was that they inspired other scientists, including physicists such as Maurice Wilkins and Francis Crick, to focus on biological problems. The intersection of physics and biology spurred a great revolution that yielded molecular biology and molecular genetics. The next great step was the determination of the double helix structure of DNA by Crick and biologist James Watson. That discovery provided the molecular basis for the great advances of the next decades.

More recently, Lynn Margulis and Dorion Sagan (2000) contributed an interesting book with the same title, *What Is Life?* However, they more or less avoided a direct answer to the question. They wrote, "The question 'What is life?' is . . . a linguistic trap. To answer according to the rules of grammar, we must supply a noun, a thing. But life on Earth is more like a verb. It repairs, maintains, re-creates, and outdoes itself." They then elaborated on the notion of life as a verb by describing the activities that characterize life. Others have used a similar strategy, and it may be the best possible answer that we can get. The fact is that life is very hard to define. It's far easier to explain what it *does* rather than what it *is* (Table 2.1). From that we can form an opinion of what life is and, in that way, understand how death lacks those qualities.

Others have also tried their hand at defining life. Stuart Kaufman (2004, 654– 665) attempted to ground the definition of life in a more biophysical or thermodynamic manner. He suggested that a living organism is "an autonomous agent in a self-replicating system that is able to perform at least one thermodynamic work cycle." It was a valiant try, but Kaufman's explanation founders when he considers how cells organize themselves: "I can only stumble with ordinary English: the cell achieves a propagating organization of building, work, and constraining organization that completes itself by formation of a second cell. Is this matter alone, energy alone, entropy alone, or information alone? No. Do we have

Table 2.1 Characteristics of Living Organisms

They regulate their internal environment (homeostasis).

They are made up of one or more cells (organization).

They use energy (metabolism).

They grow (growth).

They change in response to the environment (adaptation).

They respond to external stimuli (response).

They reproduce themselves (reproduction).

a formulated concept for what I just described? No." In many ways, his answer is similar to that of Margulis and Sagan: life is a collection of activities.

The difficulty in defining life was noted by Daniel Koshland (2002):

What is the definition of life? I remember a conference of the scientific elite that sought to answer that question. Is an enzyme alive? Is a virus alive? Is a cell alive? After many hours of launching promising balloons that defined life in a sentence, followed by equally conclusive punctures of these balloons, a solution seemed at hand: "The ability to reproduce—that is the essential characteristic of life," said one statesman of science. Everyone nodded in agreement that the essentials of a life were the ability to reproduce, until one small voice was heard. "Then one rabbit is dead. Two rabbits—a male and female—are alive but either one alone is dead." At that point, we all became convinced that although everyone knows what life is, there is no simple definition of life.

Loike and Pollack (2019) summarized life as "the property of an organism that possesses any genetic code that allows for reproduction, natural selection, and individual mortality." This definition leaves out robots using artificial intelligence, but it is open to life-forms other than those known on Earth. That is a key aspect to remember. Carl Sagan warned against the tendency to define life according to what we have learned about life on Earth—it could be quite different elsewhere in the universe.

In summary, many outstanding biologists, such as those noted here, have found it to be surprisingly hard to define life. In effect, many of their explanations could be covered by adapting Supreme Court justice Potter Stewart's famous comment about pornography: we may not be able to define life, but we know it when we see it. Most biologists would likely agree that life has several consistent characteristics that are needed for an organism to be considered to be alive. The term *autopoiesis* refers to the characteristic of living systems that allows them to maintain and to reproduce itself. It was first suggested by Chilean biologists Humberto Maturana and Francisco Varela (1980). Perhaps this is the most reasonable description of what it means to be alive.

Recognizing Life and Death

As Sagan warned, the characteristics associated with living organisms so far really only apply to those that live on Earth. This biased view ignores the obvious confounding factor: that life elsewhere in the universe might not look much like it does here on Earth. What would life look like on another planet? Just as intriguingly, what did life look like when it first emerged here on Earth? If we created a new form of life, what would it look like? Those questions are critical issues for three groups of scientists: astrobiologists, who search for extraterrestrial life; those who study the origins of life on Earth; and those who study synthetic biology.

Life in Other Places

Is there life elsewhere in the universe? If so, what form might it take? Would it be similar to life on Earth? Would it be based on carbon? What would characterize that life? Most of us here on Earth can speculate about those questions, but they are a practical matter for exobiologists.

In 2003, the first of five successful NASA rovers landed on Mars to conduct various scientific measurements. Some were aimed at looking for evidence of life (e.g., organic molecules) rather than life itself. Still, knowing what to look for requires an understanding of what is alive. In fact, NASA defines life as a "self-sustaining chemical system capable of Darwinian evolution" (NASA 2020). Steven Benner (2010) carefully reviewed the various definitions and characteristics of life and noted, "Astrobiologists are committed to studying life in the Cosmos, the terran life we know as well as the extraterran life we do not know but hope to encounter. But what exactly do we seek?" That is a great and haunting question. Certainly, writers and filmmakers have imagined many possibilities. The Mos Eisley cantina scene in *Star Wars: Episode IV—A New Hope* was widely praised for its many imaginative creatures. However, a more careful look shows that it is still heavily influenced by what we know about our own vertebrate biology.

NASA clearly has a deep interest in defining life. Its scientists and other exobiologists are searching the universe for signs of life, even if that life is not quite like life on Earth. Its website notes that the definition of life has "a distinctly gray and fuzzy quality," rather than being a bright line. That's not much help, but given how much trouble we have in defining life on Earth that is right in front of us, it's probably a pretty accurate assessment. Cleland and Chyba (2002) argue that any attempt to define life is premature until a more careful and general description of living creatures is available, perhaps including those from other worlds. That is also not a very satisfying answer, but it may be the best we have.

Origin of Life (and Death)

Research into possible life on other planets is fascinating, but the study of life right here on Earth is equally amazing. How did life begin? Interestingly, questions of life and death occur only where there is life. Without life, there is no death, but how did it all get started? *Abiogenesis* refers to the transformation of matter from nonliving to living, and how this happened is one of the great remaining questions in biology. Life might have originated somewhere else in the universe and been transported to Earth. That theory, called panspermia, has been around for some time and was revived most recently by Hoyle and Wickramasinghe (1977). Some support for panspermia has come from recent findings of biological material in meteorites. However, the only life that we are aware of in the universe right now is on Earth. If astrobiologists are successful, the concept of life originating somewhere else might need to be revisited. Even then, the original question about life on Earth would remain.

The Earth is a bit over 4.5 billion years old, and for almost a billion years after it formed, there was no life. Obviously, there was also no death. Then about 3.7 billion years ago, life appeared. The exact mechanism is not yet understood, but one of the first suggestions came from Charles Darwin in a letter to Joseph Hooker in 1871. Darwin wrote about "some warm little pond" that had a mixture of chemicals and a source of energy. He speculated that in this pond a protein substance might have been created that could change over time.

Many researchers have tested more sophisticated versions of Darwin's "warm little pond" and the recipe he suggested in broad terms. The first, in 1953, were the classic experiments by Stanley Miller and Harold Urey. They showed that basic organic chemicals resulted from the exposure of an early Earth atmosphere to electrical discharges that simulated lightning (Miller and Urey 1959). Many variations on these experiments have clearly demonstrated that various useful biological molecules result from these mixtures.

The next challenge is to find a molecule to preserve this information. Although most forms of life today store their genetic information in DNA, many scientists believe that the earliest life used RNA for this purpose. In addition to preserving information so that life can continue, RNA can act as an enzyme itself without any protein. This was a revolutionary finding. Until then, enzymes, biological molecules that catalyze chemical reaction, were all thought to be proteins. These two characteristics strongly implicate RNA in the beginnings of life. This is often referred to as the "RNA world hypothesis." But where did the RNA come from? An excellent study in 2009 partially answered this question. It showed that the basic building blocks of RNA can be assembled from simple organic molecules, such as hydrogen sulfide and hydrogen cyanide, other chemical compounds that are widely assumed to have been part of the early Earth, and ultraviolet light (Powner, Gerland, and Sutherland 2009).

Thus, at some point, this nascent chemical biology began to take on some of those characteristics of life that we noted earlier. Through the use of RNA, they could preserve the information for their own existence. This view gained additional support when Thomas Cech and Sidney Altman showed that RNA can also perform some enzymatic reactions (Cech, 1986). The early life began to tap and use energy sources to keep the reactions going. Eventually, they built membranes to contain their own internal environment. Cells resulted, and then they began to reproduce themselves. Life had begun.

And death followed. Obviously, without life, there could be no death. Those early "deaths" might have been as difficult to define as death is today. We can speculate that some of those miniature experiments in forming life failed. Some "small warm ponds" were overwhelmed by a rainstorm, flood, or high tide, or they dried up completely. Some ran out of energy before they could sustain themselves. This is not exactly the same as death today. In that early context, death was the failure of life. Like the evolution of life from self-sustaining chemical reactions to cells, death also evolved to what it is today. Defining those reactions is critical to understanding how life originated and evolved.

Synthetic Biology

One approach to determining how life could have started on Earth is to try to replicate the original experiment. In other words, is it be possible to assemble a completely artificial cell that exhibits the characteristics and activities of natural cells? A number of scientists believe that it is.

One strategy is that used by those studying the origins of life: start at the beginning with a few chemicals. Some of these issues might be solved by those studying the origin of life on Earth. For example, Zwicker et al. (2017) showed that chemical reactions in small droplets, provided an external source of energy, could show cycles of growth and division much like living cells.

Another approach is to make a cell. Making an artificial cell is a major task, but multiple laboratories have started the first steps in the project. Even the most stripped-down bacterium would still have a very large number of genes, proteins, and other molecules (Szostak, Bartel, and Luisi 2001). These would not be representative of the beginnings of life on Earth. So an initial cell might include a minimal genome of DNA or RNA in a membrane-bound vesicle, but later many more genes would have to be controlled to turn on or off in an appropriate manner (Noireaux, Maeda, and Libchaber 2011). The exchange of nutrients and production of energy are critical elements. All of these would require membranes that could support those various activities. Finally, replication would be a major challenge. Beyond these issues, there would be a myriad of others. To qualify as "living," a cell must be able to replicate and be subject to Darwinian evolution. Progress is being made on these issues too. For example, a much longer completely artificial genome was synthesized and transplanted into cells of the bacterium *Mycoplasma mycoides* as the cells' only DNA, and the resulting cells had the ability to self-replicate (Gibson et al. 2010). In a similar experiment, an artificial DNA of 100 kb was successfully used in mouse cells (Mitchell et al. 2019).

Of course, these studies also raise serious ethical issues. Will the artificial cells be considered living cells? Even more alarming are reports on the behavior of stem cells in culture (Rivron et al. 2018). Mouse and human stem cells in culture begin to automatically take on three-dimensional characteristics of embryos. How far should these experiments be allowed to proceed? Some scientists advocate consideration of these issues in advance (Piotrowska 2019). These issues are likely to be the subject of future discussions.

Could a new type of organism be constructed from existing cells? Synthetic organoids can be assembled from different cell types, but they cannot be controlled by the experimenter. Groups of individual cells in a petri dish sometimes assemble into structures. For example, cardiac cells can make connections and begin to beat in synchrony by themselves.

In a truly astounding series of experiments, Kriegman et al. (2020) report the translation of computer-generated structures into "living" things, which they term *xenobots*. Cells taken from the African clawed frog (*Xenopus laevis*) self-organize into a life-form that can live for about a week. Pluripotent cells and various progenitor cells were combined, with minimal manipulation, into structures designed for a simple biological function, such as movement. Their food source is bits of yolk that ordinarily would aid embryonic development. They lack nervous systems, but they can repair themselves and, most amazingly, they can perform simple life functions. These things (it's hard to come up with an appropriate descriptor) clearly challenge any definition of what it means to be living.

Bacteria with Synthetic DNA

While defining life has been a daunting challenge, the one seeming constant in living organisms on Earth is that all use DNA and/or RNA to encode their genetic information, and that DNA or RNA code is based on combinations of four nucleotides (adenine, guanine, cytosine, and thymine, for which the symbols are A, G, C, and T).

But now even that one constant has been lost. In 2011, a group of scientists led by Steven Benner (Yang et al. 2011) added another complication. They developed two more nucleotides (isoguanine (B) and 1-methylcytosine (S)) and incorporated them into DNA. In 2019, they added two more (5-aza-7-deazaguanine (P) and 6-amino-5-nitropyridin-2-one (Z)), increasing the number to eight

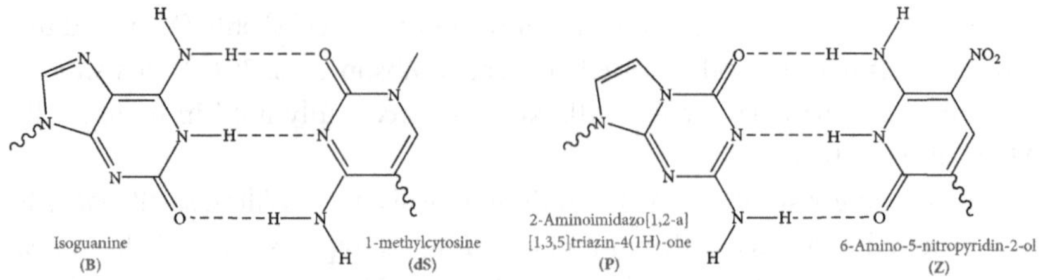

Figure 2.1 Hachimoji DNA bases. This simple structural formula depicts hydrogen bonding between the unnatural bases in hachimoji DNA, as described by Hoshika et al. 2019.

(Hoshika et al. 2019). The new nucleotides are similar to the original purines and pyrimidines, and they use hydrogen bonding to create base pairs (S with B and P with Z). The new DNA is called hachimoji DNA (*hachimoji* means "eight" in Japanese) (Figure 2.1). The extra nucleotides will increase the amount of information that can be stored and provide additional chemical and physical properties to the DNA. The DNA was transcribed into RNA with no problems. Even more amazingly, Zhang et al. (2017) created a semisynthetic bacterium that uses six nucleotides. The modified *E. coli* bacterium is stable and seemingly healthy. It can import the unnatural nucleotides. Importantly, it seems to store information indefinitely using the new nucleotides.

These experiments read like science fiction. They also show that living organisms can fulfill all of the characteristics of life even without the traditional four-nucleotide genetic code that all life on Earth uses. The additional base pairs in synthetic DNA provide these bacteria with the ability to encode more information. Some have speculated that such organisms might have an evolutionary advantage over natural life-forms, and that advantage might raise ethical questions about these semisynthetic organisms.

At the Boundary of Living and Nonliving

Poe envisioned a shadowy and vague boundary between life and death. Since then, literature, television, and the movies have been remarkably inventive in imagining many creatures that inhabit that twilight region. Since its publication in 1897, Bram Stoker's *Dracula* has begat countless other vampires. Zombies walked in George Romero's cult classic *Night of the Living Dead*. The undead now inhabit many shows on television. While these are all entertaining, even more amazing creatures actually exist in nature, and many of those seem to exist at the boundary of life and death.

Dormancy

Some accomplish this trick of straddling the boundary between life and death by being dormant for extended periods. When they are dormant, the processes and activities associated with life are so minimal that the organism may seem dead. For some, the only way to determine that they are indeed alive is to attempt to shock them out of their dormancy and see whether they begin to grow again.

Bacterial Endospores

When some bacteria encounter conditions in which food is scarce, they protect themselves by forming endospores. For bacteria of the genus *Firmicutes*, the endospores retain only the bare minimum of structures and functions required for life. And they can live for amazingly long times. For example, Cano and Borucki (1995) found spores that seemed to be related to a strain of bacteria called *Lysinibacillus sphaericus* (formerly *Bacillus sphaericus*) in a bee preserved in amber for the last 25–30 million years. Those spores were revived and started growing. More amazingly, using extremely stringent isolation procedures to prevent contamination, Vreeland, Rosenzweig, and Powers (2000) isolated bacterial spores from a salt deposit that formed 250 million years ago and found them to be still viable. Those salt deposits formed around the time that trilobites were still alive and before the dinosaurs appeared. The characterizations of living organisms that were detailed in the preceding sections would be essentially impossible to detect in these undisturbed spores. In fact, the only way to show that these bacteria are alive is to test them for growth in culture.

Fungal Spores

Fungi do not use spores to survive stressful conditions. Instead, spores are a reproductive mechanism. No spores have been found that can match the longevity of the bacterial endospores, but these spores can live for quite a long time. For example, potato wart disease, which is caused by the fungus *Synchytrium endobioticum*, was found in a field in Pennsylvania thought to be disease free for more than twenty-five years (Putnam and Sindermann 1994). As with bacterial endospores, the longevity of the fungal spores defies the criteria normally associated with being alive. But clearly, they are not dead.

Seeds

Seeds can also remain dormant for long times. As one example, seeds of a date palm were found at the Israeli city Masada, where the Jews committed mass suicide rather than surrender to the Romans in the year 73 or 74 CE (Sallon et al. 2008). They had lain dormant for 2,000 years but were still viable.

Frozen Embryos

To help families have children, a number of methods of in-vitro fertilization have been developed. Typically, those procedures produce more embryos than are needed to be implanted in the uterus, and the extras are frozen and stored to be used if the first efforts fail. Those techniques have been very successful. Many babies have been produced, and lots of extra embryos have been frozen. It is not clear that there is a limit to the length of time an embryo can be frozen and still result in a successful pregnancy. Embryos have been used that were frozen for more than twenty years (Dowling-Lacey et al. 2011).

Having Some Characteristics of Life

The organisms just discussed blur the distinction between life and death by lying dormant for extended periods. Other organisms have some of the characteristics of being alive, but not all. The organisms living at the boundary between living and dead are at least as mysterious as the zombies of literature and the movies. These "undead" but not quite living species include viruses, viroids, and prions. As one indicator that they are at the boundary of life, there is considerable disagreement as to whether they should be considered living or not.

Viruses

Viruses are among the most ubiquitous entities on earth. The first virus was discovered in the 1890s: tobacco mosaic virus, which was found to remain in solution after the solution had been filtered to remove bacteria-sized pathogens. It is remarkably stable. It can remain dormant for over fifty years and still infect plants (Scholthof 2000). Since the discovery of tobacco mosaic virus, thousands of viruses have been found. They infect animals, plants, fungi, and bacteria, and some are parasitic on other viruses. Many more remain undiscovered.

Viruses exist at the edge of being alive. In fact, most scientists have traditionally classified them as nonliving. They lack several of the characteristics noted previously that are usually associated with being alive. For example, viruses do not metabolize anything. These parasites can reproduce only by hijacking the cellular systems of their host. Viruses are usually far smaller than other living cells, and they encode a very small number of proteins. Importantly, they cannot live outside of a cellular host: they are obligate parasites. They do not produce their own energy, instead commandeering the host cell's machinery to produce proteins and nucleic acids. In other words, they reproduce, and that is just about it. And to do even that, they need help. Thus, the idea that viruses are nonliving seemed to be reasonable.

Of course, viruses are not the only obligate parasites; there are others (Zomorodipour and Andersson 1999). The life cycle of the bacterium *Chlamydia trachomatis* features two forms. An elementary body is the infectious form that allows transmission to other hosts. A reticulate body replicates by binary fission. The bacterium *Rickettsia prowazekii*, which causes typhus in humans, replicates sometimes in a cell's cytoplasm and sometimes in the nucleus. Both of these bacteria have lost large portions of their original genomes, and now they lack the ability to carry on some of the normal functions of free-living organisms. As obligate endoparasites, they cannot survive without the help of their host. They have more genes than most viruses, and they are much larger. However, they differ from viruses only in degree and thus muddy the waters of whether viruses are alive or not.

The issue became even more complicated in 2003 with the discovery of giant viruses (La Scola et al. 2003; Van Etten 2011; Colson et al. 2020). These viruses are much larger than an ordinary virus. Some are nearly as big as a bacterium. In addition, their genomes encode an unusually large number of genes. Some have 1,000 genes. Others have up to 2,500. Most surprisingly, the giant viruses' genomes encode some of the substances needed for translation (the production of proteins from an mRNA template, the first step in independent self-replication), a characteristic never before associated with a virus (Schulz et al. 2017). Some of the genes seen in these giant viruses seem to be related to those in cellular organisms. Perhaps the viruses "borrowed" these genes—viruses are notorious for taking up and dropping off extraneous sequences from their hosts—but the answer in this case is not clear. There are other similarities between some types of viruses and living cells. The Marseilleviridae have doublets of the core histones H2A and H3, found in all eukaryotes (cells that have a nucleus and other organelles surrounded by a membrane and organized chromosomes) (Erives 2017). Needless to say, these findings have caused viruses to be seen in a new light. Hegde et al. (2009) and Villarreal and Witzany (2010) provided reasons for why viruses should be included among living organisms. While it used to be "clear" that viruses were nonliving, these new lines of evidence provide a more complex picture and provide clues to the evolution of viruses and how they fit into a larger evolutionary scenario. Forterre (2010) defined an organism as "an ensemble of integrated organs (molecular or cellular) producing individuals evolving through natural selection." That definition clearly includes viruses as living organisms.

These unusual viruses are so different from typical viruses that some scientists have suggested that they belong in a fourth domain of life, or at least that they be classified as a new order of viruses (Colson et al. 2012). They further speculate that there might be many more of these giant viruses to be found. The methods used to isolate viruses generally involve filtering out everything of a certain size on the assumption that viruses were so small that they would pass through the filters. However, these giant viruses would have been captured by the filters and so not identified as viruses.

Others still disagree that viruses are alive. Moreira and Lopez-Garcia (2009) provided ten reasons why viruses should be excluded from living organisms, and van Regenmortel (2016) noted that the claim "that viruses are alive is only a metaphor based on anthropocentric interpretations of viral replication." Just because the so-called viro-cells have compartments in which new viruses are made does not qualify them as living organisms. Navas-Castillo (2009) gives his version of six reasons why viruses should not be considered as living. Among them is the "sterile earth argument." If any virus was inoculated onto a sterile Earth, it would not grow. However, if every type of bacteria were placed on a sterile Earth, many of them would immediately grow.

The issue of whether viruses are living or nonliving is not settled and likely will not be soon. If viruses are considered to be nonliving, then the discussion ends. However, if viruses are included among living organisms, we have only kicked the can down the road to even simpler organisms.

Viroids and Satellites

Viroids and satellites are even more bizarre than viruses. The simplest viruses consist of a nucleic acid genome and a protein coat. Viroids have a genome of a short strand of circular, single-stranded RNA but completely lack a protein coat. They are pathogens of plants. Theodor Otto Diener discovered the first viroid, potato spindle tuber viroid d (PSTVd), in 1971. No viroid is known to cause human disease (Zheng et al. 2017).

Viral satellites are widespread in eukaryotic and prokaryotic cells (prokaryotic cells lack a nucleus and other organelles that are surrounded by a membrane). These subvirus particles contain a very small nucleic acid genome with the information to make their own protein coat (Krupovic and Cvirkaite-Krupovic 2011). However, they can replicate only with the help of a coinfecting virus. Thus, they are a parasite on a parasite. Most of them infect plants, but some infect mammals, arthropods, and bacteria.

Prions

Prions are poorly understood, but they seem to be misfolded proteins that are associated with some neurodegenerative diseases, including transmissible spongiform encephalopathies (mad cow disease, scrapie in sheep, kuru and Creutzfeldt-Jacob in humans). Some have suggested that Parkinson's disease may be a prion-like disease. Prions are "infectious" in that the change from a normal structure to a disease structure can be passed on from protein to protein and perhaps from cell to cell. Amazingly, no nucleic acid seems to be involved.

Life, Death, and the Law

As we have seen, biology has trouble defining both life and death, and many organisms seem to exist in a gray area at the boundary of life and death. That inability to define these terms is not really a problem for scientists. Biologists can easily live with an imprecise description of the characteristics of life. However, some situations require a more exact determination of life and death. That need is most apparent at the two extremes of life, the beginning and the end, conception and death, and those examples often involve the law. And most are painfully emotional situations.

When does human life begin? The debate around the beginning of human life is one of the most contentious ever and drives the controversy around abortion. Does life begin at conception or at some later time, perhaps when the heart begins to beat or when the fetus is viable on its own? The characteristics associated with life are not helpful in this debate. Advances in perinatal medicine (e.g., lung surfactants that allow immature lungs to work more effectively) have helped more preterm babies than ever to survive. In-vitro fertilization has enabled more couples to have children but left a tricky question of what should be done with the fertilized embryos that are not used.

The questions around the issue of when life begins are more political and philosophical than biological, but there is more than enough biology to confuse the issue. To begin with, both the ovum and the sperm are themselves living cells. So no real beginning of biological life occurs; life simply continues in another form. Given this consideration, the question is more about the character of the life. When does the joining of the egg and sperm result in the beginning of human life? Here there is plenty of room for differences of opinion. Does human life begin at conception, at implantation, when a heartbeat begins, at viability, or according to some other measure?

It is unlikely that modern medicine will slow down. Further advances are likely to push the time of viability back even further. More pre-term babies will survive, and the disagreements will continue. It is not a question for this book.

The other extreme is equally fraught. When does human life end? This question is no less vexing than the previous one, and the lack of a clear definition can result in tragic and extremely emotional situations. For example, in 1990, a twenty-six-year-old woman named Terri Schiavo suffered a cardiac arrest. Paramedics were able to resuscitate her, and at a local hospital she was intubated and artificially ventilated. Tragically, she had been without oxygen for too long and suffered massive brain damage. Even with extensive treatment and support, she remained in a coma. Eventually she was declared to be in a persistent vegetative state. Patients in a persistent vegetative state have lost their higher brain functions, which involve their personality, reasoning, and remembering. However, they retain their lower brain functions, which control heartbeat, breathing, digestion, and other essential functions, and so they can live for extended periods, sometimes even without respiration support. In 1998, her husband asked the courts in Florida to declare her brain dead so that her feeding tube could be removed and she could be allowed to die. He argued that she would not want to be kept alive artificially in such a state. Schiavo's parents opposed the action. They hoped that their daughter would eventually come out of the coma, and thus they wanted her to continue to live. Years of court maneuvering resulted. Eventually, the Florida legislature, Florida governor Jeb Bush, the US Congress, and even President George W. Bush became involved on the side of the parents. Finally, in 2005, the court agreed with her husband, the feeding tube was removed, and Schiavo died.

New Standards for New Technologies

The traditional standard for the presence of life involves the presence of a heartbeat and breathing. If no pulse or breathing can be detected, the person is dead. Breathing was often checked by holding a mirror to the nose and mouth. If the mirror did not become fogged, no breathing was detected, and the person was thought to be dead. This definition has been used for millennia, and its great advantage is that it matched the common understanding of what is living or dead. The loss of either heartbeat or breathing meant that the other would also stop within short order. Using this standard, time was not an issue. The deceased could be observed for minutes or hours to ensure that there was no spontaneous recovery of breathing or heartbeat. Even laypeople could see that a person was dead by this standard.

However, advances in modern medicine have complicated the situation, and now it can be excruciatingly difficult to make such a determination when it is needed most. Cases often result in considerable emotional pain. Disagreements among the experts exacerbate the situation. As with the Terri Schiavo case, those cases can last for years and be heartbreaking for everyone.

Ironically, the confounding advance was the invention of the mechanical ventilator or respirator. This device has saved many patients. By breathing for them, the ventilator keeps the various tissues in the body well oxygenated, and often this gives the body time to heal. However, physicians early on began to notice that they had two types of patients on respirators. One type of patient being kept alive by the machine might eventually recover. The ventilator was a godsend for those patients. The other type of patient was essentially already dead. Mollaret and Goulon (1959) called this second group of patients "coma dépassé" or beyond a coma. They would never be revived. These patients complicated the definition of death by not fitting into its traditional definition.

Over time, the concept of brain death began to be considered in the determination of death. In 1968, a committee at Harvard Medical School suggested
that loss of brain function should be used as the definition of death. That rec-
ommendation was revolutionary. Until then, the standard involved only heartbeat and breathing. Later President Jimmy Carter asked leading authorities to
resolve some of the uncertainty around the determination of death. In 1981, the
National Conference of Commissioners on Uniform State Laws, the American
Medical Association, the American Bar Association, and the President's
Commission for the Study of Ethical Problems in Medicine and Biomedical
and Behavioral Research released a report that included a model state law, the
Uniform Determination of Death Act (President's Commission 1981). It sets two
criteria for determining death: irreversible loss of circulation and respiration
or irreversible loss of all functions of the entire brain, including the brain stem.
A person meeting one or the other of these criteria would be deemed dead. While
not all states adopted all provisions of the model law, all fifty states adopted these
criteria for determining death because they provide useful measures. They give
physicians a test that helps them decide whether to discontinue expensive and
possibly futile treatment in certain patients.

In the United States, the states vary in who can declare death. One physician
is sufficient in most states. Some require confirmation by a physician training in
the neurosciences if brain death is the basic of the determination confirmation.
Alaska and Georgia allow a registered nurse to declare death with the proviso
that it be confirmed by a physician within twenty-four hours. Some states require
a confirmatory test, such as an apnea test—a test of blood gases to determine if
there is any respiratory activity. Interestingly, New Jersey has a religious exemp-
tion that allows relatives to opt out of the use of brain death as a criterion for determination of death.

Similar guidelines have been adopted in many other countries as well.
Wijdicks (2002) examined the laws of eighty countries and found that seventy
recognized brain death as part of their law, and fifty-five used brain death to
allow organ transplantation; some require additional clinical testing to support
that diagnosis. Apart from the countries examined in this study, guidelines differ,
but many countries have some regulations regarding the determination of death
if not transplantation. So far, China and much of Africa have none.

In 2012, the World Health Organization and the Canadian Blood Services
cosponsored a meeting of experts in Montreal to review standards for the determination of death (WHO 2012). They noted that any standard must take into
account philosophical, religious, and cultural differences surrounding death,
the lack of scientific knowledge about the process of death, disagreements about
methods of determining death, and the profound emotions surrounding death.
While they viewed their work as preliminary to further study, they advised

against using the death of any organ as a standard for death. Rather, they focused on the loss of function: "Death occurs when there is permanent loss of capacity for consciousness and loss of all brainstem functions. This may result from permanent cessation of circulation and/or after catastrophic brain injury. In the context of death determination, 'permanent' refers to loss of function that cannot resume spontaneously and will not be restored through intervention."

To determine brain death, physicians use a checklist of items. Obviously, they check for a pulse and for breathing. In addition, they check the patient's reflexes, such as blinking, coughing, and gagging if the back of the throat is touched. They test for involuntary responses, such as dilation of pupils when exposed to a bright light. In total brain death, all functions of the brain and brain stem cease. Circulation and breathing must be maintained artificially. Total brain death is determined by a number of tests, including electroencephalographs, physical tests, and various forms of imaging. In some cases where the clinical examination is unclear, a confirmatory test after a determination of brain death is performed. These tests reexamine electrical function in the brain and cerebral blood flow. The requirement for them is not universal, and the practice varies from country to country. Wijdicks (2010) noted that confirmatory tests are generally not needed and may provide information that might confound the determination rather than clarify it.

Today there are two major approaches to defining death (Holland 2017). The first focuses on biology. Under this approach, life is determined by function of the cardiorespiratory system. If the person has no heartbeat and no respiration, they are dead. Alternatively, the person is dead if they have lost the function of the entire brain or brain stem. The second approach involves consciousness (Sarbey 2016). This method is more subjective and focuses on higher brain functions rather than the entire brain. Those who support this definition see those characteristics that make us individuals, including memory, consciousness, and personality, as the most important. Advocates for this view point out that there is a difference in our perception of the death of a human and that of a plant or animal. For them, the death of specific parts of the brain, particularly the cerebrum, is the key. Under this definition, death of the whole brain is sufficient for death but not necessary. The loss of the cerebrum and those higher brain functions are all that is needed. Of course, this standard is more difficult to measure clinically.

Even with such progress in setting the criteria for determining death, complicated and tragic cases occur in which family members and even medical experts disagree. One such case occurred in Oakland, California, in 2013. While recovering from routine surgery, thirteen-year-old Jahi McMath suddenly suffered a massive hemorrhage that deprived her brain of oxygen. She was declared brain dead, but mechanical ventilation kept her "breathing." Her parents insisted she was alive and asked a court to invalidate her death certificate. They argued to the court that since her heart was still beating, she was alive. The case was heartbreaking and resulted in a great deal of acrimony. In 2014, the family transferred her to a facility in New Jersey that recognizes personal and religious objections to a declaration of brain death. Disagreements by medical experts continued to complicate the case. In 2018, Jahi suffered kidney and liver failure, and life support was ended, allowing her to die.

Organ Donation

Edgar Allan Poe scared generations of people with his story about premature burial. As we've seen, premature burial was somewhat of an urban legend at the time, and in any case, it is far less likely nowadays. If he were writing today, Poe might use premature organ harvesting as his premise.

Organ failure is a major cause of death. Kidneys, livers, lungs, and hearts become diseased or wear out. In the past, this inevitably meant death, but in 1954, Joseph Murray performed the world's first organ transplant when he removed a kidney from a healthy twin and placed it in the other twin, who was dying of kidney failure. In 1961, Dr. Murray also performed the first transplant of a kidney from a cadaver to a living patient. These successes won him the Nobel Prize in Medicine and started a whole field of transplant medicine. Today, hearts, lungs, kidneys, livers, and corneas are routinely transplanted. Indeed, transplants have saved many lives since then. However, the demand for healthy organs far outstrips the supply. Every day, ten new patients are added to the US transplant waiting list; in 2017, seventeen people died every day for want of a transplant.

A healthy donor can easily give up one of their two kidneys to save the life of a sick person. However, clearly, a heart cannot be donated by a living person. Thus, transplants from cadavers represent a rich supply of potential organs. And this raises another complication for medical ethics. Death is not simultaneous across all organs and tissues. Kidneys and livers are fairly robust, but hearts are notoriously sensitive to oxygen deprivation and begin to deteriorate rapidly after death. Organs for transplant must be harvested as soon as possible after a person has died, and so time has become an issue and pushed the boundaries of the criteria for determining death.

Ethically, physicians must do nothing to hurry the death of a person. A person must be declared dead before organs can be removed for transplant. And not only do organs die at different rates, but so do people. This can make the difference between healthy organs for transplant and not. The complication for transplants is that the precious minutes are ticking along as the transplant team waits for permission to begin the harvest and organs begin to deteriorate. This obvious conflict of interest is partially solved by forbidding the transplant team to have any input at all into the determination of death.

As noted earlier, determining death is not easy, and determining death by loss of circulatory function has been controversial. Some believe that artificial means of perfusing the organs (e.g., chest compressions) prevent the death of the brain stem, which means death does not occur while someone is being artificially ventilated. In this view, death depends on a conscious decision by caregivers to terminate support. This complication in the definition of death is felt especially by those who want to harvest organs for donation (Hanto and Veatch 2010).

There are several categories of loss of brain function. First, a person in a coma is profoundly unconscious—that is, they are never conscious. They do have relatively normal sleep-wake cycles and respond to certain stimuli. When they wake up from the coma, they retain most or even all of their faculties.

Second, individuals such as Terri Schiavo who have been unconscious for four weeks and meet certain other criteria are declared to be in a persistent vegetative state (Beecher et al. 1968). A persistent vegetative state is not the same as a coma. It involves the loss of higher brain functions that control thinking, memory, and personality. The person will never regain those. However, the lower brain functions remain intact and continue to control the person's breathing, body temperature, heart rate, and other basic functions. Occasionally, patients may seem to partially wake up; they may have some spontaneous eye or body movement, may make sounds such as cries or laughter, or may respond to external stimuli. These are just random events and not conscious organized activities. However, relatives and friends may see more in those movements than is warranted and take false hope from that. Patients have been known to recover from a persistent vegetative state, but it is rare, and after three months it is extremely rare. Another tragic example is that of New Jersey resident Karen Ann Quinlan, who in 1975 suffered severe brain damage after drinking alcohol and taking Valium. She lost consciousness and stopped breathing for fifteen to twenty minutes; eventually she lapsed into a persistent vegetative state. Her EEG showed only slow-wave activity, and even that was abnormal. Slow wave activity is part of the non-REM (non-rapid eye movement) component of normal sleep. She was fed by a nasogastric tube and needed a ventilator to breathe. For religious reasons, her parents asked the hospital to turn off the ventilator to allow Quinlan to die. Under threat from the local prosecutor that he would prosecute them for murder, the hospital and the Quinlans went to court. The court agreed with them, but when the ventilator was turned off, Quinlan continued breathing on her own for nine more years. She eventually died in 1985.

Third, patients who have lost all brain function are in dire straits. They have lost even the ability to maintain basic bodily functions, such as breathing. If the respirator is turned off, they cannot breathe on their own, and they will die.

Finally, another rare disorder has some similarities to coma, persistent vegetative state, or total brain death but is not the same. Patients with locked-in syndrome are completely paralyzed except for eye movement. The causes include traumatic brain injury, circulatory diseases, neurological diseases that involve loss of the myelin sheath that protects nerve cells, and medication overdoses. The patients are completely conscious and can think, but they cannot speak or move. There is no cure, and only some patients recover some function.

Interestingly, for people who fall into one of the first three of these four categories and are declared brain dead, physicians have gone back to a version of the cardiopulmonary criteria for a final determination of death. Patients in any of these categories are considered to be dying but not dead. It is in these cases that individual rights to control one's own body sometimes compete with societal needs for transplantable organs.

In 1997, the Institute of Medicine (now the National Academy of Medicine) was asked to come up with guidelines for determining death in brain-dead patients so their organs could be transplanted. They recommended two conditions. First, the transplant team can have nothing to do with the decision by the family to suspend life support. The decision has to be made by the family in consultation with their own physician. Second, the patient has to be carefully monitored for five minutes after life support is terminated in order to detect any spontaneous heart activity or breathing. If any signs of life are noted, the patient must be returned to life support and the transplant is canceled. These rules are referred to as "donation after cardiac death." These rules have been in use for some time now, and brain-dead patients who have experienced cardiac death now contribute a significant fraction of the organs that are transplanted.

Yet even with these new rules, the pool of potential cadaver donors is still far too small to fill the need. Those five minutes for observation do not seem like much, but they can mean the difference between healthy organs and useless ones. They are useless because the organs begin to deteriorate shortly after they lose their supply of oxygen. The original five-minute rule was somewhat arbitrary. It was based not on any research but simply on expert opinion. Could the time for observation be shortened in any way? The speculations are not as heartless as they might seem at first. Many parents of small children who have been declared brain dead actually want to donate the child's organs. Perhaps in that way they could see some good coming from the loss of their child, or could imagine some part of their child continuing to live. In any case, many of those parents support a loosening of the rules.

Some logical questions arise from this protocol. If the person is going to die anyway, does the time allotted for monitoring them matter? Why not leave them on life support, remove the organs, and then turn off life support? Isn't there something wrong with waiting for a heart to stop in one patient so it can be restarted in another patient? But if a patient remained on life support, could one really contend that that heart failed? In light of these issues, a number of compromises have been made. For example, the arbitrary five minutes has been reduced to sixty-five seconds at some institutions. In other cases, large intravenous lines are placed in veins in the patient's groin before life support is interrupted. These lines allow the heart and other organs to be quickly flushed with fluids to extend their useful life once death has been declared.

James L. Bernat (2013) provides an excellent review of the problems associated with determining death. Organ donation is driving the need for a clear determination of death. The dead donor rule states that the patient must be declared dead before organs are harvested. However, controversies remain with using brain death to define death. Should it be the whole brain or the brain stem? How many examinations should be required? In certain unusual circumstances, such as therapeutic hypothermia (a procedure in which the temperature of a patient suffering cardiac arrest is significantly lowered to give the heart and other organs time to recover), when should those tests be attempted? Similarly, controversies remain with determining death by loss of circulation. How long should asystole (total lack of electrical activity in the heart, or "flatlining") last before a person can be declared dead? Is the loss of circulation permanent? The one thing everyone agrees with is that those who are determining death must be completely separate from those involved in organ harvesting.

"Gap Between Body and Death"

The issue of determining death continues to evolve as medical technology advances. For the vast majority of cases, the traditional cardiopulmonary criteria are adequate. For some difficult cases, the brain criteria are useful. Still, modern medicine leaves us with a "gap between body and death," as Ben Sarbey (2016) put it in an excellent paper. He also suggests that any future guideline focus on those aspects of humans that we value most: the higher functions of personality and memory. Unfortunately, current technology offers only poor means of measuring these functions. In addition, measuring these functions opens new issues. How will we view patients with significant dementia? Are they alive or dead? Shah, Truog, and Miller (2011) examine what they describe as the "legal fictions" of determining death in terms of total brain failure. While they maintain that those with total brain failure sometimes regain some degree of function, their principal objection is the lack of transparency in a process that is driven mainly by the need to secure organs for transplantation. The one thing that is clear is that the controversy is not likely to be settled soon.

Where Next?

Science continues to move forward. Boundaries will be pushed and passed. Those advances may further clarify the definitions of life and death and the tests for death. However, they are equally likely to further confound the issues. For example, a recent experiment may present challenges to our definitions of death. Researchers showed that some functions of pig brains that had been dead for up to four hours could be reestablished (Vrselja 2019; Vrselja et al. 2019). The brains were connected to pumps that circulated a cell-free solution that contained a hemoglobin-based oxygen carrier and other chemicals. The brains showed no electrical activity: the artificial blood used to nourish the brains contained nerve blockers. However, the researchers observed "preservation of cytoarchitecture; attenuation of cell death; and restoration of vascular dilatory and glial inflammatory responses, spontaneous synaptic activity, and active cerebral metabolism in the absence of global electrocorticographic activity."

While this experiment might call to mind Mary Shelley's book *Frankenstein, or the Modern Prometheus*, in which Dr. Frankenstein brings to life a creature made from dead tissues, the ability to revive brain activity in dead individuals will not be realized in the near future, if ever. However, these results raise lots of ethical questions and have possible implications for future practices with research animals and for the removal of organs from patients after they are declared dead (Farahany, Greely, and Giattino 2019; Younger and Hyun 2019).

3

What Kills Humans

> Behind every man now alive stand 30 ghosts, for that is the ratio by
> which the dead outnumber the living.
>
> —Arthur C. Clarke

Approximately 7.5 billion people are alive in the world today. That's a lot of people, but as Arthur C. Clarke observed, many more people lived in the past. In fact, more than 108 billion people have lived since modern humans appeared on the Earth. Mr. Clarke's estimate might be a little off by current numbers, but his point is clear. A lot of people lived before us, and most of them have died.

An estimate, such as this one, requires some assumptions, and to make this estimate of 108 billion, the Population Research Bureau (PRB) in Washington, DC, made several (Kaneda and Haub 2020). Various hominid (human-like) species existed over the last couple of million years, and any estimate depends when you start counting. Kaneda and Haub limited themselves to modern humans and so set 50,000 years ago as the start date. Written records are available for only about the last 10 percent of that time, but there are well-accepted methods for making estimates for earlier humans. The world population was about 5 million at 8,000 BCE. The growth rate was low because death was common. Many died in infancy, and for much of our history, life expectancy at birth was about ten years. Life expectancy is a statistical measure of the average time a person is expected to live, based on their year of birth, current age, and sex. Life expectancy at birth is the average time a newborn is expected to live. By accounting for all of these factors, the PRB came to an estimate of 108 billion people. So Clarke's estimate was actually quite good. It is even better if we remember that he made it when the Earth's population was only about 3 billion.

So more than 100 billion people have lived and died on Earth. What happened to them? Why did they die? Trauma and injuries? Diseases? This chapter will explore why humans die, then and now. The major causes are pretty much the same, but the relative importance of the major causes has changed over time. Even today, the reasons people die differ in developed and developing countries and between the rich and poor within countries. We will begin with the current

The Biology of Death. Gary C. Howard, Oxford University Press. © Oxford University Press 2021.
DOI: 10.1093/oso/9780190687724.003.0003

situation and work our way backward to the beginning of humans. Finally, we will look to what might be the future.

Mosquitos and Human Death

In his book *The Mosquito: A Human History of Our Deadliest Predator*, Timothy Winegard (2019) proposes an intriguing hypothesis. He asserts that the lowly mosquito and the diseases it spreads have caused the deaths of nearly half of the people who have ever lived on Earth. He made this estimate by extrapolating from the annual deaths caused by mosquitos. Deaths from malaria and other mosquito-borne diseases (e.g., dengue fever, yellow fever, chikungunya, West Nile, Zika, lymphatic filariasis, eastern equine encephalitis, tularemia) come to about 2 million each year (other groups estimate about 700,000 deaths annually). This is far more than from any other animal. By comparison, snakes kill about 50,000 people each year and dogs about 25,000. The mosquito family includes about 3,500 species. They are found throughout much of the tropical and temperate regions of the Earth, and their ranges are increasing with global warming. Even if Winegard's estimates are only half correct, it is still a very large number—and an interesting idea.

Figure 3.1 Mosquito (*Aedes aegypti*). Photograph courtesy of the Centers for Disease Control and Prevention.

Dying Today

Natural Causes

"During the time that I'm speaking this sentence, a dozen people just died, worldwide. *There*. Another dozen people have perished" (Freitas 2002). In his paper, Robert Freitas compared what we typically call natural death to a human holocaust. The annual numbers are staggering. In 2001, nearly 9 million people in India died of natural causes, another 9 million in China, and over 2 million in the United States. That year nearly 55 million people died of natural causes worldwide (Freitas 2002). Put another way, about 733 of every 100,000 people alive at the start of 2001 had died of natural causes by the end of that year. With the term *holocaust*, Dr. Freitas may be overstating the case a bit—the world now has over 7.5 billion people, and because India, China, and the United States are three of the most populous nations on Earth, they would be expected to have the lion's share of deaths.

Natural causes is the most common reason given for death. The term is often used to describe the death of older persons in particular. One of those was Dwight Eisenhower. On the morning of March 28, 1969, he died at Walter Reed Army Medical Center in Washington, DC, at age seventy-eight, of natural causes. More specifically, the cause of his death was congestive heart failure. Ike had experienced a lot in his life. He survived the stress of being the supreme commander of the Allies in World War II while living on little more than coffee and cigarettes. He smoked three to four packs of cigarettes a day. He also survived two terms as president, but just barely—he suffered a heart attack and stroke and endured Crohn's disease while in office. After retiring, he had six more heart attacks. But in the end, he succumbed in 1969 to natural causes.

Typically, the term *natural causes* is typically used to describe the death of an older person. Children and younger people may die of natural causes too, but we rarely note that fact. In normal conversation, we often use it to describe the loss of an elderly person. We might say, "She lived a full life." Or "He died in the fullness of his years." We may be sad about their passing, but we are not surprised. In the "normal" order of things, the old are expected to die.

Coroners and medical examiners use the term with more specificity. To them, any death caused by old age or disease is natural. The person died of an internal disease process (e.g., diabetes, cardiac arrest, Alzheimer's disease) or simply because their body had deteriorated due to old age. Unnatural deaths include homicide, suicide, and accidents. Some fall into a gray area between two categories, and the final designation on the death certificate is at the discretion of the coroner. Nevertheless, as Freitas notes, natural causes take a terrible toll on humanity every day.

When and Where Do People Die?

It's an old cliché that everyone dies. But where? Few people think about where they might die. Death comes in a seemingly random manner, at least as far as most of us can tell. Few people want to die in a sterile institutional environment, such as a hospital. They want to be in a familiar supportive place surrounded by their family and friends. Most people (about 80 percent) would like to die at home, but only about 20 percent do (Broad et al. 2013).

We might guess that most people die in a hospital. Using death certificates from New York State in 1977, Katz, Zdeb, and Therriault (1979) found that about 60 percent of people died in hospitals. Males of all ages were more likely than females to die in a hospital, and older people were more likely to die in a hospital than younger people. Many people also died in nursing homes. Females were particularly more likely to die in a nursing home. Just 27 percent of all deaths occurred outside of some institution. Broad et al. (2013) examined 16 million deaths in hospitals, residential care facilities, and other places since 2001. Overall, half occurred in hospitals, although the percentages varied by country. For example, in Japan, 78 percent died in a hospital, but in China, only 20 percent did. Overall, about 18 percent of deaths occurred in a residential center, but the percentages doubled for about every ten additional years of age and were 40 percent higher for women.

Interestingly, these numbers might be changing. Emergency department visits are down, and the number of deaths in the emergency room declined by nearly half from 1997 to 2011 (Kanzaria, Probst, and Hsia 2016). A study of deaths among Medicare recipients from 2000 to 2015 (Teno et al. 2018) found that the likelihood of dying in a hospital was greatly reduced. Several factors might be involved (Shmerling 2018). Patients might be considering institutions other than hospitals for end-of-life care. Many people now value quality of life over length of life. They do not want heroic measures taken to prolong life that lacks quality. In fact, quite a number of patients now understand that in some cases further treatment is useless. Many physicians agree. Nevertheless, hospitals remain a primary location for deaths. Sometimes insurance companies will only pay for care in a hospital. In other cases, there just isn't anywhere else for a very sick patient to go. Finally, doctors are trained to do everything they can rather than to step back and let things simply proceed.

One might also wonder when most deaths occur. Are there particular times of the day, or a month or season, that pose a greater risk to health? It turns out that there are. For example, the number of deaths spikes around the Christmas and New Year's holidays. Perhaps influenza, colder weather, and the stress of the holidays combine to put additional pressure on people. Researchers in New Zealand tested that idea (Knight et al. 2016). Christmas for them takes place in summer, and so they could directly test the possibility that cold weather caused the extra deaths. However, they found increased deaths even during their warm holidays. Thus it cannot be the cold weather or seasonal diseases that kill people in greater numbers around holidays. Others have speculated that the stress of the holidays is responsible for the deaths. Mondays also show an increase in the risk of cardiac death (one might be tempted to attribute that to *Monday Night Football*). Witte et al. (2005) showed that the risk was increased for both men and women. Changes in the circadian rhythms of medication, work and leisure time, or other factors are suspected to be involved.

Time of day is also a risk factor. A study of nearly 5,000 death certificates in New York City for 1979 found a significant increase in the risk of death beginning at 2:00 a.m. and peaking at 8:00 a.m., with a second peak at 6:00 p.m. (Mitler et al. 1987). Three factors might contribute to the peaks. First, people might die at any time in the night, but they are only discovered when others wake up in the early morning. Medical care might have been less available in the early hours. Finally, there might be an increase in risk associated with the circadian rhythms of humans. After a more detailed study, they found that only the deaths of people over sixty-five showed an increased risk in the early morning; deaths in those under sixty-five did not. Different diseases did or did not show spikes over the course of the day, depending on disease and gender (e.g., ischemic heart disease, 8:00 a.m.; hypertensive disease in women, 1:00 a.m.; cerebrovascular disease in men, 6:00 a.m.). The results are interesting, but it is not clear that any definite conclusions can be made.

The *Bible* says, "For a man also knoweth not his time" (Ecclesiastes 9:12). Doctors routinely give some patients, such as those with terminal cancer, an estimate of their life expectancy. One way these estimates are used is to plan future care; for example, those patients with a life expectancy less than some specified amount of time (e.g., six months) can qualify for hospice care. However, a new biomarker reported by Deelen et al. (2019) may give clinicians a better understanding of who is most vulnerable, even among those who do not have a specific life-threatening disease. Their examination of over 44,000 patients aged 18–109 used a profile of fourteen metabolites to generate a score for risk of mortality five to ten years out. For one cohort, the score was 83 percent accurate, whereas the traditional scores, based on systolic blood pressure and total cholesterol, were only 78 percent accurate.

Death in the United States

Today many of the health issues that plagued humans for centuries have been mostly overcome, at least in developed countries. Over the last century, life expectancy has increased dramatically. According to the US Centers for Disease Control and Prevention (CDC), life expectancy at birth for both sexes and all races was 47.3 years in 1900. In 1950, it was 68.2 years. In 2000, it was 76.9 years. In the last century or so, parents have known that their children would likely live to grow up and that the parents will probably die before their children. However, we now experience a new set of diseases that are associated with long life (e.g., heart disease, cancer, Alzheimer's disease). From 2001 to 2014 in the United States, income mattered for life expectancy (Chetty et al. 2016). Those with higher incomes tended to live longer. Those in the top 1 percent of income lived longer than those in the bottom 1 percent (fifteen years longer in men and ten years in women). And for the last few years, life expectancy has gone down in the United States, particularly among less educated, blue-collar workers in the Midwest and rural South.

Emerging viruses have become a much more visible threat (Morens and Fauci 2013). At this writing, the world is reeling from the 2020 coronavirus pandemic, but in recent years, we have seen several other viruses appear or reappear. These include other coronaviruses (e.g., MERS, SARS), arboviruses (e.g., Zika virus [Fauci and Morens 2016]), and Ebola and Marburg viruses. All these are spreading due to human activity (such as urbanization and international travel). There also remains the threat of a resurgence of a more virulent version of influenza. With climate change, disease vectors are moving. Southern Europe, for example, may be more susceptible to epidemics (Hotez 2016).

Each year the CDC provides the numbers around deaths in the United States (e.g., CDC 2020c) (Figure 3.2). This "butcher's bill" makes for grim reading. For example, 2,813,503 people died in the United States in 2017 (Kochanek et al. 2019). The causes vary, and they have changed over time. In the United States and much of the developed world, the incidence of infectious diseases has declined with advances in medical sciences and the relative availability of antibiotics and other drugs. But now we are experiencing a new set of diseases that are associated with long life. Heart disease and cancer have topped the list of causes of death in the United States for some time; together they account for half of all deaths among Americans. Alzheimer's disease is a major cause of death in the United States, and as the population continues to age, the incidence of this and other neurodegenerative diseases is increasing rapidly. "Diseases of affluence" (such as obesity, diabetes, and heart disease) are also on the rise (Ezzati et al. 2005).

Genetics can play a role in the things that we die from. For example, more than 3,500 known diseases follow Mendelian genetics (Brunham and Hayden 2013)—that is, they are the product of a single defective gene. Among the first to be described was Huntington's disease, an autosomal dominant neurodegenerative disease that usually manifests late in life. Other conditions

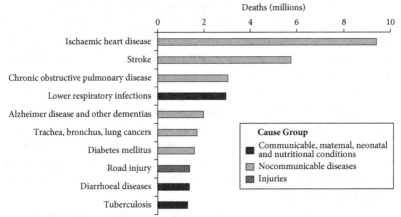

Source: Global Health Estimates 2016: Deaths by Cause, Age, Sex, by Country and by Region, 2000–2016.
Geneva, World Health Organization; 2018.

Figure 3.2 Top ten causes of global death. *Source: Global Health Estimates 2016: Deaths by Cause, Age, Sex, by Country and by Region, 2000–2016* (Geneva: World Health Organization, 2018).

are more complicated—they might involve interactions of multiple genes or genes and the environment. Our health and death are significantly affected by our genetic makeup, the environment that we live in, and interactions between them.

The human body is a remarkable machine that requires careful regulation of multiple systems. Death comes from the loss of critical bodily functions, but what causes that loss of function is specific for each disease. For example, death might result from a lack of oxygen, but that lack can be the product of loss of circulation, loss of blood, the lungs' inability to absorb oxygen, shock, the inability of the brain to organize bodily systems, or other reasons. For many diseases, the actual mechanism through which the disease causes death is not fully clear; environmental and genetic factors may play a role.

Heart Disease

Heart disease is the leading cause of death in the United States and in many other countries. In the United States, over 600,000 people die of heart disease each year. The major risk factors are high blood pressure, high cholesterol levels, and smoking. Other risk factors include diabetes, obesity, poor diet, physical inactivity, and alcohol.

Different types of heart disease are associated with the different functions of the heart, though those functions are interrelated. The most common type of heart disease is coronary artery disease, in which deposits of cholesterol and other material (known as plaques) build up in the walls of the arteries of the heart. Eventually, the deposits may grow large enough to partially block the artery and reduce the supply of oxygen and nutrients to a part of the heart muscle. The process is called atherosclerosis, and over time, the heart muscle may become weaker, resulting in heart failure or arrhythmia (irregular heartbeat). In some cases, a vessel can become completely occluded, or a plaque may rupture and block the artery. When that happens, the part of the heart muscle supplied with oxygen and nutrients by that artery begins to die. If enough of the heart's pumping ability is affected, the brain gets inadequate supplies of oxygen-carrying blood. Death results when the brain runs out of oxygen, and that happens within minutes.

The beating of the heart is caused by an electrical stimulus that causes the heart muscle to contract. It is controlled by a group of specialized cells in the sinoatrial (SA) node in the right atrium (right upper chamber of the heart). The electrical impulse from these cells causes the heart muscle to contract in a steady, rhythmic fashion. However, if those cells fail, the heart has other cells that will cause the heart to beat too fast, too slow, or irregularly. In severe cases of cardiac arrhythmia, the contractions become so irregular that the heart loses its ability to pump blood at all. The electrical cells of the SA node can be damaged in a variety of ways, including a lack of oxygen, drugs, congenital conditions. In those cases, the pathways themselves can be damaged, and the cells of the myocardium can fail to respond to the impulses.

Heart failure occurs when the heart cannot pump enough blood to supply the body's needs. The main diseases contributing to heart failure are conditions that damage the heart, such as ischemic heart disease (which occurs when a part of the heart is not receiving sufficient blood supply), high blood pressure, and diabetes. Nearly 6 million Americans suffer from heart failure, and about half of those diagnosed die within five years.

Cancer

Although heart disease is the number one killer in the United States today, people probably fear cancer more than heart disease. However, many cancers are now treatable, and survival rates continue to improve for most cancers.

The National Cancer Institute estimated in 2018 that over 1.7 million new cases of cancer would be diagnosed that year, and that over 600,000 people would die from cancer (NCI 2018). About 38 percent of all men and women will be diagnosed with cancer at some point in their life. The most common cancers are breast cancer, lung cancer, prostate cancer, colorectal cancer, melanoma, bladder cancer, non-Hodgkin's lymphoma, kidney and renal pelvis cancer, endometrial cancer, leukemia, pancreatic cancer, thyroid cancer, and liver cancer. The most lethal cancers are lung cancer, colorectal cancer, breast cancer, pancreatic cancer, prostate cancer, leukemia, non-Hodgkin's lymphoma, liver cancer, ovarian cancer, and esophageal cancer.

Although there are many types of cancers, they do have a set of common characteristics. In most normal tissues and organs, cells grow until that tissue or organ is at its appropriate size and shape. For example, most adult human livers weigh about 3 pounds or so and have a typical "liver" shape. Those characteristics are controlled genetically. Once the appropriate size and shape have been achieved, the cells of that tissue or organ exit the cell cycle—they no longer divide. There is some cell division to replace worn-out cells, but all of that is carefully controlled. There are some tissues that are regularly worn out by use—skin, hair, intestinal lining, blood cells—and those tissues continue to have cell division to replace lost cells. Still, the key concept here is that all of this growth (or lack of it) is carefully regulated.

Cancer cells have lost this careful regulation of their growth and begin to divide again. Some cancers grow in place; this type is generally referred to as benign. However, benign tumors are not completely problem-free—their increased size may encroach on a neighboring tissue in a way that can cause trouble (for example, cancers in the skull can put pressure on the brain). In other cancers, some cells may break off from the original mass and be carried by the blood system to another site in the body, in a process called metastasis; this type of cancer is referred to as malignant. Cells may initiate tumors in other organs, such as the lung or liver, that will interfere with the functions of those tissues.

Not all cancers form a solid tumor. Some involve blood cells, and there are several types of these illnesses. Leukemias (such as acute lymphoblastic leukemia, acute myeloid leukemia, and chronic lymphocytic leukemia) are cancers of blood-forming tissues, such as bone marrow. Lymphomas (such as non-Hodgkin's lymphoma) typically affect the lymph nodes, and myelomas cause too many white cells of a particular type to be produced. Leukemias may have few or no symptoms, or they may cause symptoms such as fatigue, weight loss, infections, and abnormal bleeding and bruising. Lymphomas may cause symptoms, such as enlarged lymph nodes, fever, and weight loss.

So how do cancers kill? As a general statement, as cancers grow, they take up more space or invade other tissues. Thus, they can put pressure on the other organs and interfere with their functions. For example, cancers can damage blood vessels and cause dangerous bleeding.

About half of cancer patients suffer from cachexia or "wasting" (Martignoni, Kunze, and Friess 2003), which causes them to lose weight due to a loss of fat and muscle. The causes are unknown, but they are thought to involve the release of toxic factors into the blood by the tumor, including inflammatory cytokines. The wasting makes the side effects of chemotherapy more serious, reduces the patient's quality of life, and increases the risk of death.

Because cancer occurs in such a wide variety of tissues and organs, its effects on the body also differ. For example, cancer might block the digestive tract, so food cannot be absorbed by the intestines. As a consequence, other organs and tissues will be deprived of vital nutrients and energy. Cancer might cause part of a lung to be blocked, resulting in an infection. A as the tumor spreads, the amount of effective lung tissue will be reduced, and the patient will slowly suffocate from lack of oxygen. Bone cancer can disrupt the calcium balance of the body by causing the release of calcium from bones into the blood; when the amount of calcium in the blood is too high, it can be fatal. Bones are also the site of blood cell production, and cancer can reduce the ability to replace worn-out blood cells—without adequate red blood cells, the patient will not be able to transport enough oxygen to vital tissues, such as the brain. The loss of platelets might result in abnormal bleeding; without white cells, the patient will be more vulnerable to infections. The liver is a complex organ with multiple functions in the body, including the regulation of multiple salts and chemicals, and disruptions of these can be fatal. The pancreas also produces many hormones and enzymes, and cancer in it can be particularly troublesome.

In cases of metastasis, cancers can occur in multiple organs and tissues all at once, and the effects can be additive. These cancers are more difficult to treat because it is hard to determine if all of the sites of metastasis have been eliminated by treatment.

Fortunately, great progress has been made. Excellent treatments are available for many types of cancer, and five-year survival rates are up for many types of cancers. Among the most exciting new treatments are those that enable our own immune system to attack cancers, such as melanoma, which is particularly dangerous (Redman, Gibney, and Atkins 2016).

Diabetes Mellitus

Diabetes is a group of diseases that result from the inability to control the levels of glucose (or sugar) in the blood. Glucose is the primary source of energy for cells, and the protein insulin enables cells to take up glucose from the bloodstream. Insulin is secreted by beta cells in the islets of Langerhans, structures that are found in the pancreas. If insufficient insulin is made or if the insulin cannot be captured by receptors on the surface of cells throughout the body, glucose levels in the blood become too high for extended periods.

There are three types of diabetes. In type 1 diabetes, an autoimmune condition causes the destruction of the beta cells that make the insulin (in some cases, the cause of that destruction is unknown). Type 1 is sometimes called juvenile onset, but it can occur at any time (though it is still relatively rare to see it develop in older people). In type 2 diabetes, the body loses its ability to use insulin effectively. This condition, called insulin resistance, is normally associated with obesity, a lack of exercise, poor diet, and stress, and typically occurs around middle age. However, the poor diet and exercise regime of younger children in the Western world has put them at great risk for developing type 2 diabetes at a much younger age. Sadly, the increase in type 2 diabetes among those ten to nineteen years old parallels the increase in severe obesity in that age group (Laffel and Svoren n.d.). Finally, gestational diabetes occurs in some pregnant women; after the woman gives birth, it usually ends. According to the CDC, 34.2 million people (10.5 percent of the US population) have diabetes, and 88 million adults have prediabetes (34.5 percent of the adult population) (CDC 2020a).

Diabetes is a debilitating disease that slowly attacks many systems. The risks of coronary artery disease, stroke, and peripheral artery disease are greater for diabetics than for nondiabetics. Damage to small blood vessels affects eyes (causing blindness), kidneys (resulting in the need for dialysis or transplants), and nerves (leading to foot ulcers that sometimes necessitate amputation). Most diabetics suffer from fatty deposits in their arteries; the result is atherosclerosis (hardening of the arteries) around the heart and a two to four times higher risk of stroke.

Accidents (Unintentional Injuries)

Accidental deaths include those from burns, drowning, falls, traffic accidents, and poisoning (deaths from diseases, homicides, and suicides are not included). According to the CDC, there were 169,936 deaths from unintentional injuries in the United States in 2017. This made accidents the third-leading cause of death. Those accidental deaths included over 36,000 from falls, 40,000 from traffic accidents, and nearly 65,000 poisonings. By comparison, homicides accounted for over 19,000 deaths, and three-quarters of those were by firearms. Suicides cost another 23,000 lives, and nearly half were by firearms. Accidental deaths are even more important for infants through teenagers (up to nineteen years old) than for adults; over 12,000 of them each year, and the rate of accidental death for males is nearly twice that of females. The cause varied by age: suffocation was responsible for two-thirds of deaths in children under one year, and drowning was the leading cause for those ages one through four, and traffic accidents for those five through nineteen.

Chronic Lower Respiratory Disease

Chronic lower respiratory disease, or CLRD (formerly called chronic obstructive pulmonary disease), includes three diseases: chronic bronchitis, emphysema, and asthma. In bronchitis, the bronchial tubes in the lungs are inflamed and swell, making breathing difficult. The cause is usually viral, but sometimes it is caused by bacteria. In emphysema, the structure of the small airways called the bronchioles is changed. In a healthy person, tissue around the bronchioles holds them open so that air can be expelled. Loss of that tissue in emphysema means that the patient cannot empty their lungs; in turn, they cannot breathe in oxygen-rich fresh air. Asthma is a chronic condition that involves inflammation and narrowing of the airways. These actions make breathing difficult and cause wheezing, tightness of the chest, shortness of breath, and coughing. More than 16 million people in the United States have CLRD, and others have probably not been diagnosed. In these diseases, the patient gradually loses lung surface area until there is not enough for effective gas exchange. Death often comes from respiratory failure—the lungs simply can no longer work, and the body is deprived of oxygen. The patients eventually succumb to what amounts to suffocation.

Other lung-related causes of death include heart failure, pneumonia and other infections of the lungs, and pulmonary embolisms.

Stroke (Cerebrovascular Disease)

According to the CDC, nearly 140,000 Americans die from a stroke each year. African Americans have nearly twice the risk of a stroke as white Americans.

A stroke occurs when the blood flow to the brain is interrupted. There are two types of strokes. In an ischemic stroke, the blood supply is reduced by a blockage in the artery, similar to a heart attack. In a hemorrhagic stroke, the artery ruptures, and blood flows into the brain. Different areas of the brain have different functions, and the death of cells in specific areas results in the loss of those functions. If critical areas are killed, the patient might die.

Brain tissue is very sensitive to oxygen levels and will begin to die if blood flow is not restored within minutes. If the interruption lasts longer, larger areas will begin to die. The lack of oxygen stops the production of adenosine triphosphate (ATP), which is essential for numerous functions within the cells. But the cell continues to consume ATP, and rapidly uses up the available supply. Once that happens, the ion pumps that transport calcium ions out of the cell can no longer function; as a result, glutamate is released, the process of apoptosis (programmed cell death) is initiated, and ultimately the cell dies.

Influenza and Pneumonia

Influenza, a disease caused by a virus, occurs in annual outbreaks. Each year, 3–5 million people catch the flu, and 250,000–500,000 die worldwide. Periodically, very large outbreaks, called pandemics, kill many more. In 1918, the Spanish influenza killed 50 million. In 1957, the Asian influenza killed 2 million. In 1968, the Hong Kong influenza killed 1 million. In 2009, H1N1 influenza (swine flu) killed 100,000–400,000. The very high mutation rate of the influenza virus allows it to change rapidly and has hampered efforts to produce a single vaccine that controls all variants of influenza.

Interestingly, it isn't always the virus that actually kills the patient (Jabr 2017). The virus enters the body through the mouth or nose or even the eyes. It replicates and spreads rapidly, and this elicits a massive immune response. The immune cells attack and destroy the infected cells. In most healthy people, the immune system can defeat the virus. However, some people are more vulnerable, such as the very young, the very old, and those with preexisting medical conditions. In them, the release of cytokines and chemokines from the dying infected cells can cause significant tissue damage. For example, so many lung cells may be destroyed that the patient can no longer obtain sufficient oxygen. In other cases, someone weakened by an influenza infection may develop a bacterial infection, such as streptococcus or staphylococcus, and that secondary infection might cause a huge immune response. Alternatively, the bacteria might enter the blood system and cause an overwhelming toxic shock that causes multiple organs to shut down.

Pneumonia can be caused by an infection of the lungs by viruses or bacteria (and sometimes by other organisms), by medications, or by autoimmune diseases. The accompanying inflammation affects the ability of the small air sacs called alveoli to absorb oxygen. Pneumonia is serious mostly in the elderly, the very young, and those with other health conditions. It is a leading cause of death around the world. The famous physician William Osler, who practiced in the latter part of the nineteenth century and the early part of the twentieth, once called pneumonia the "friend of the aged"—a way for the sick to die relatively painlessly.

COVID-19

Around the beginning of 2020, a new disease suddenly appeared. COVID-19 is caused by a novel coronavirus called SARS-CoV-2. It seems to have originated in China, but it rapidly spread around the world. In less than a year, more than 500,000 Americans died (CDC 2020b), and many more died around the world. It is now the third-leading cause of death in the United States, behind only heart disease and cancer. The virus seems to kill by at least two mechanisms. First, the virus attacks lung cells and renders them incapable of exchanging oxygen. Second, it also causes blood clots that can be lethal. The course of the disease varies greatly. Some infected people show very few, if any, symptoms. Those with preexisting conditions (e.g., respiratory or heart problems, diabetes, obesity) often have a worse course.

Nephritis, Nephrotic Syndrome, and Nephrosis

Kidneys remove waste products from the blood; regulate levels of water, salt, and blood pressure; and help to make red blood cells. For example, the kidneys work

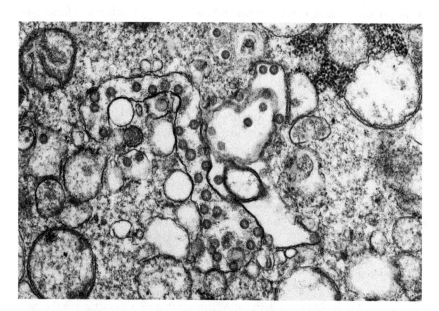

Figure 3.3 SARS-CoV-2. This virus is the cause of COVID-19. Transmission electron microscopic image of an isolate from the first US case of COVID-19. The spherical viral particles, colorized blue, contain cross-sections through the viral genome, seen as black dots. Image obtained by Hannah A. Bullock and Azaibi Tamin of the Centers for Disease Control and Prevention.

with several other organ systems to control blood pressure by regulating the levels of sodium and potassium in the blood. The kidneys make an enzyme called renin. If blood volume or sodium levels are too low or potassium levels too high, the kidney releases renin. The renin acts on angiotensinogen, which is made in the liver, to change it to angiotensin I. Angiotensin-converting enzyme in the lungs changes angiotensin I to angiotensin II, which causes blood vessels to constrict to raise blood pressure. Angiotensin II also causes the adrenal glands to release aldosterone, which causes the renal tubules to retain sodium and water and release potassium. Any disruption of these vital functions can be problematic. Kidneys can be damaged by disease (e.g., infection or diabetes), high blood pressure, or injury.

Kidney function can be diminished by bacterial infection, by an autoimmune reaction that attacks the kidneys, or by strenuous exercise that results in damaged blood cells. When loss of function is caused by infection, the condition is called nephritis; when the cause is noninfectious, it's called nephrotic syndrome or nephrosis.

The main result of loss of kidney function is the retention of toxins in the blood. For example, urea is a metabolite of amino acids or proteins. Creatinine results from the breakdown of muscle tissue. Normally, these toxic metabolites and many others are eliminated by the kidneys in the urine. At high levels in the blood, they damage many organs and disrupt many biochemical functions. Without kidney function, death is likely within two to three weeks.

Intentional Self-Harm (Suicide)

According to the CDC, 47,000 Americans killed themselves in 2017. It was the tenth-leading cause of death overall, but the second-leading for those ages eighteen to thirty-four. More women attempt suicide than men, but men succeed more often. Sadly, the rate of suicide is increasing dramatically: the annual US suicide rate increased 24 percent between 1999 and 2014. Worse still, suicide carries a lot of stigma, and the rates may be underestimated. Firearms account for nearly half of all suicides. Suffocation and poisoning round out the top three causes.

Alzheimer's Disease

Alzheimer's disease (AD) is a devastating disease that robs its victims of the very characteristics that define us as human: memories and personality. Most patients with AD survive for years while those aspects slowly degenerate. Patients over the age of sixty-five generally live four to eight years after diagnosis with AD. Its cause is unknown.

Although AD is a terrible disease, it does not directly kill patients. However, the decline in brain function that accompanies the disease and the inevitable dementia have many side effects. Patients cannot accomplish even minor daily activities. For example, swallowing becomes difficult, and patients often aspirate food into their trachea, resulting in aspirational pneumonia. Bedsores, sepsis, and infections in general, injuries, and starvation and dehydration also lead to death.

The population of the United States and many other developed countries around the world is aging. Since age is a major risk factor for AD and other neurodegenerative diseases, the outlook is grim.

Most people who come down with AD are at least sixty-five years old. After age sixty-five, the risk of AD doubles every five years. After age eighty-five, nearly one-third of people have AD. In 2019, about 2 million people with AD were eighty-five years old or older (Alzheimer's Association 2019). The total number of people with AD and other dementias is expected to rise dramatically in the next years, because the baby boomers are rapidly reaching the age of AD risk. By 2050, 7 million people will be eighty-five or older and will have AD. Worse still, in the more than 100 years since Dr. Alzheimer first described the disease, we have not been able to develop any effective treatments. In addition, other neurodegenerative diseases, such as Parkinson's disease, amyotrophic lateral sclerosis, multiple sclerosis, and frontotemporal dementia, are expected to increase.

Genetic Diseases

Genetics is a strong factor in many diseases and suspected in others. Among the first diseases to be recognized as genetic was sickle cell anemia. Sickle cell anemia occurs mostly among populations from or with ancestry in sub-Saharan Africa. The disease results from a single nucleotide change in the gene for the production of beta-globin, which forms part of hemoglobin, the iron-containing protein that carries oxygen in the blood. More specifically, a change of GAG to GTG encodes a valine at position 6 instead of a glutamic acid. Most such changes in the nucleotide sequence have no effect, but this one does. Erythrocytes or red blood cells carry oxygen in the blood. The hemoglobin in those cells that binds the oxygen consists of two beta-globin protein molecules. The mutation that causes sickle cell disease causes a change in the three-dimensional structure of beta-globin such that the cells change shape from a disc to a sickle form. Cells with the sickle form do not flow smoothly in small blood vessels, and they also tend to clump together. The substitution can affect one or both of the gene copies that all of us have for beta-globin, and it is inherited as an autosomal recessive gene. If only one copy is affected, the person may suffer from a lack of oxygen under stress conditions, such as high altitude. If both copies are affected, the symptoms are much more severe.

How can a gene that is detrimental to survival be retained in the population? Shouldn't it be slowly eliminated, since those with the disease gene are likely to show poorer survival rates and thus to reproduce less? The answer is that there must be some positive evolutionary pressure to retain the defective gene. In fact, there is. Those with the "defective" sickle cell gene have greater resistance to malaria. On balance, the ability to resist malaria is beneficial to those who live in areas where malaria is endemic. With other diseases, the benefit is not clear. The cause of Huntington's disease is well known. It might remain in the population because the disease manifests well after the reproductive years. However, the apolipoprotein E4 isoform is linked to Alzheimer's disease and poor recovery from brain injuries, some forms of heart disease, and more, but its benefit is unknown.

Modern genetics, completion of the Human Genome Project (which mapped the entire human genome), and exciting new technology such as the gene-editing CRISPR technique are helping with the discovery of more genes connected to disease. For example, the ability to treat breast cancer was enhanced by the discovery of the two genes called BRCA1 and 2. The genes encode proteins that repair DNA breaks and slow tumor growth. Mutations in these two genes do not directly cause breast cancer; rather, women with a BRCA1 mutation have a 55–65 percent chance of developing breast cancer before age seventy, and women with a BRCA2 mutation have a 45 percent chance. Other genes have similarly complicated benefits and detriments. For example, the apolipoprotein E4 isomer (apoE4) is the greatest known genetic risk factor for Alzheimer's disease and is also responsible for problems with cholesterol metabolism. The positive pressure to retain the gene that produces apoE4 is not clear. It's also not fully understood how the presence of apoE4 increases risk for AD. Robert Mahley's laboratory has suggested several possibilities (Mahley, 2017), including toxic fragments of apoE4 and damage to mitochondria. In any case, the development of Alzheimer's disease seems to involve a first insult (an injury of some kind). Thus, head trauma such as a concussion could provide the first insult that sets in motion a series of events that result in AD.

Death in the World

The World Health Organization (WHO) estimates that 56.9 million people died in 2016. That year the population of the world was estimated as a little over 7 billion. Thus, about 0.8 percent of the world's population died that year.

They died from many different causes. WHO lists the leading causes as ischemic heart disease, stroke, and lung diseases (Figure 3.1). These causes cover deaths in both developed and developing regions. Some are more important in developing countries, such as HIV/AIDS and diarrheal diseases. Others occur at high rates in both developing and developed countries, such as diabetes and heart disease.

The risk of death from a disease differs from county to country. For example, in the United States, esophageal cancer results in about four deaths per 100,000 people. However, in Pakistan, the death rate is 140 per 100,000. The difference is likely due to the fact that chewing tobacco and drinking very hot liquids are more common in Pakistan than in the United States. Many factors in specific countries and regions combine to yield the particular set of diseases that are prevalent in that country. These include drought, flooding, exceptional cold spells, war, unstable governments, and even intentional genocide.

For the last fifteen years, ischemic heart disease and stroke have remained the world's biggest killers, with 15.2 million deaths in 2016 (WHO 2020). In other areas, there have been some recent changes in the major killers. Death from dementias doubled between 2000 and 2016, making this now the fifth-highest cause of death. HIV/AIDS fell out of the world's top ten causes of death

Table 3.1 Causes of Deaths in the United States in 2017

Disease	Deaths
Heart disease	647,457
Cancer	599,108
Accidents (unintentional injuries)	169,936
Chronic lower respiratory diseases	160,201
Stroke (cerebrovascular diseases)	146,383
Alzheimer's disease	212,404
Diabetes	83,564
Influenza and pneumonia	55,672
Nephritis, nephrotic syndrome, and nephrosis	50,633
Intentional self-harm (suicide)	47,173

Source: Heron M (2019) Deaths: Leading causes for 2017.
National Vital Statistics Reports 68: 6.
https://www.cdc.gov/nchs/fastats/leading-causes-of-death.htm

(2000: 1.5 million deaths; 2016: 1.0 million). Road injuries (2016: 1.4 million deaths) seem to be a male way of death (men: 74 percent of deaths).

The Earth contains a lot of people, and the WHO divides them into high-, medium-, and low-income countries for the purpose of determining disease and death trends. Lower respiratory infections caused many deaths across all income groups. The causes of death differ considerably between the high- and low-income groups. Over half of the deaths in low-income countries are caused by so-called Group 1 conditions: communicable diseases, nutritional deficiencies, and problems with pregnancy and childbirth. Noncommunicable diseases cause most death in all countries, but all but one of the top ten causes of death in high-income countries were noncommunicable diseases. Low-income countries also had the greatest number of deaths from road accidents.

The National Academies (NA n.d.) notes that a small number of infectious diseases kill large numbers of people each year, especially in the developing world. Many of these have ravaged humans for thousands of years, but antibiotics, vaccines, and other advances in modern medicine and public health have controlled these diseases in the developed world. These include infections of the lower respiratory tract (e.g., pneumonia), diarrheal diseases (especially in small children), HIV, tuberculosis, and malaria. HIV and tuberculosis are closely linked.

What Kills Children Today

For millennia, childbirth has been a very dangerous proposition for both mother and child. Even now, nearly 3 million women and newborns die each year (UNICEF 2020). Mothers die from complications such as high blood pressure during pregnancy (preeclampsia, which causes damage to the kidneys and liver), bleeding, or infections. The first twenty-eight days are the most critical for neonates. In 2018, about 7,000 newborn babies died every day. Although efforts to save children have seen some success, the decline in newborn deaths lags behind that of older children. Most newborns die because they are born before they have fully developed; others have a fatal birth defect or they contract an infection. Deaths also vary around the world. In the least-developed countries in sub-Saharan Africa, women are nearly fifty times more likely to die in childbirth than women in more developed countries, and their babies are ten times more likely to die in their first month.

In the United States, accidents are the leading cause of death in children and adolescents, according to the CDC. Other causes of death vary by the age of the child. For those from birth to one year, the causes are mostly related to pregnancy and childbirth. They include genetic conditions present at birth, problems due to premature birth, and the health of the mother in pregnancy. The most serious birth defects include heart defects, lungs that are not fully developed, and neural tube closure problems that might result in anencephaly. From the ages of one to four years, the most common causes of death are accidents, genetic defects, and homicide. For those five to fifteen years of age, the causes are accidents, cancer, and—shockingly—suicide.

According to the WHO (2019), the leading causes of death in children worldwide are prematurity, pneumonia, birth asphyxia and trauma, infections, congenital abnormalities, diarrhea, sepsis, injuries, malaria, noncommunicable diseases, meningitis, measles, HIV, and tetanus. Neonatal deaths (defined by the WHO as the first 27 days of life) accounted for 56 percent of child deaths. The good news is that mortality in children under five has decreased by 56 percent from 1990 to 2016, and neonatal mortality has decreased by 49 percent. Diarrhea is a leading cause of death in children under five years of age. It kills more children than AIDS, malaria, and measles put together (Wardlaw et al. 2010). In 2008, nearly 9 million children under five years of age died of pneumonia or diarrhea.

Noncommunicable diseases are a huge problem in the children and teenagers. The majority of these result from childhood conditions and behaviors. These include smoking, lack of physical exercise, heavy consumption of alcohol, and obesity.

What Do We Believe Kills Us?

Unfortunately, most people are not very good judges of risk, and the media do not always help. The twenty-four-hour news cycles that focus on mass shootings, terrorist attacks, and other dramatic killings detract from our ability to make good judgments. The last couple of decades have seen a number of tragic mass shootings, including at schools. As horrible as these events are, they are also rare. Psychologists refer to the tendency to value the latest information more than previous information as recency bias. A dramatic event seen on the news can seem like more of a threat than it really is.

For example, fear of flying is one of the most common phobias. People tend to fear flying much more than travel by car. However, the chances of dying in a car accident are far greater than the chances of dying in a plane crash. The National Safety Council (NSC 2020) puts the odds of dying in a plane crash at 1 in 9,821. The odds of dying in a car crash are 1 in 114, making it eighty-six times more likely than dying in a plane crash. Furthermore, the commercial aviation industry in the United States has a remarkable safety record. In April 2018, an engine exploded on a Southwest Airlines jet. A passenger was killed by the flying shrapnel. He was the first passenger killed in

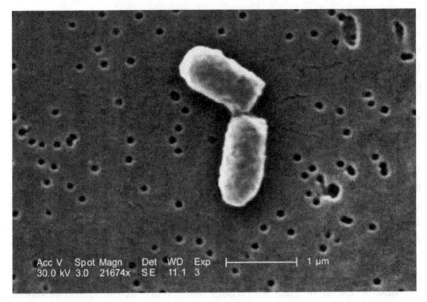

Figure 3.4 Escherichia coli. Under high magnification (21,674X), this digitally colorized scanning electron microscopic image depicts a view of a dividing, *E. coli* bacterium, clearly displaying the point at which the bacteria's cell wall is splitting into two separate organisms. Image obtained by Evangeline Sowers and Janice Haney Carr of the Centers for Disease Control and Prevention.

an accident involving a US commercial airliner since 2009. Between those two dates, nearly 100 million flights had carried several billion passengers safely to their destinations.

People tend to overestimate the odds of terrible events even though they are rare and to underestimate the odds of more common threats. For example, we buy organic foods to avoid pesticide residues, but then use pesticides in our backyards. Emotions weigh more in our decision-making than facts. We fear a nuclear plant although the risk of a meltdown is extremely low. More-hazardous activities that receive less attention slip out of our minds. For example, taking children to fast-food restaurants is far riskier in the long run than many things that scare people.

Diabetes, obesity, smoking, and heart disease are quiet killers. There is nothing dramatic about them except the damage that they cause. However, with these conditions the harm typically occurs in slow motion, and it garners very little attention.

Death in the Distant Past

In *The Leviathan* (1651), English philosopher Thomas Hobbes famously characterized human life as "solitary, poor, nasty, brutish and short." That was certainly the case for prehistoric humans—life was short and death rates were high.

Modern humans have existed for about 50,000 years, but our hominid ancestors lived hundreds of thousands or even a couple of million years before that. For most of that time, hominids and then humans were hunter-gathers in small nomadic communities, with only rare contact with other groups. We can gain some appreciation of those challenges by studying populations that largely still live this way and have had very little exposure to modern cultures. Those groups—there are presently about 100 of them around the world—subsist much as they have for thousands of years. They live in various places around the world: the Andamanese on islands in the Bay of Bengal; the Toromona and Pacahuara in Bolivia; the Ayoreo in Bolivia and Paraguay; the Awá, Kawahiva, and Korubu in Brazil; the Nukak in Colombia; the Tagaeri and Taromenane in Ecuador; and other groups in Peru, Venezuela, and New Guinea.

We might also extrapolate from our closest relative, chimpanzees, in the wild. Their environment is similar to that occupied by our distant ancestors. Anderson (2018) found that the chimps are susceptible to a number of diseases, including pneumonia and respiratory viruses. Some diseases cause death directly. Others do not kill but weaken the animal so that it becomes susceptible to accidents or predation. Lions and leopards kill chimpanzees; hyenas and African hunting dogs generally leave healthy chimps alone but will attack weak, sick, or old ones. Snakes are also potential threats: pythons take chimps, and venomous snakes may kill others. Falling from trees is a significant threat to chimpanzees. Falls often result from fights. Drowning is another problem. And attacks by other chimps occur; sometimes the losing chimp is eaten. All of these might have occurred with our human ancestors.

Like the chimpanzees, early humans were undoubtedly experts in surviving in their environment. They knew how to deal with the various challenges from nature to find food, water, and shelter. Trauma from injuries and infections must have taken a toll. Malnutrition and even outright starvation were likely common, and infectious disease must have killed many. Mothers died in childbirth, and many infants did not survive long. Life span was quite short, perhaps twenty-five to thirty years.

Remains of prehistoric humans have been found in many areas of the world. The causes of death can sometimes be determined by examining those remains. While the soft tissue is often lost, the marks of wounds left on a bone might be attributable to an arrow or knife. Teeth can also give hints about health and disease.

Carbon dating showing that a number of people of the same age died around the same time in a specific area might indicate a disease outbreak. In some cases, even soft tissue has survived, preserved by freezing, by drying, or in bogs.

Schultz (1999) examined bone samples from several fossil hominids and highlighted the examples of two Neandertal men. One had suffered a broken arm that had healed but showed bone atrophy, likely due to the inability to use the arm after the bone healed. The other was determined to be a more mature individual who displayed signs of osteoporosis.

Bioarchaeological efforts have documented a number of diseases and conditions that afflicted early humans. Interestingly, most of these are diseases that still afflict people today. Of course, we now have far better treatments, and most of these can be controlled.

Tuberculosis

Humans have suffered with tuberculosis for thousands of years. The presence of disease can be inferred from the morphology of bones in human remains; clues include new bone deposited near the site of infection, spondylitis (inflammation) of the lower thoracic and upper lumbar vertebrae, and osteomyelitis (infection in the bone) that can result in significant loss of bone tissue. The earliest known possible case of tuberculosis was found in a skeleton of *Homo erectus* in Turkey from the Middle Pleistocene, around 490,000–510,000 years ago (Kappelman et al. 2008). Numerous other cases of likely tuberculosis have been reported.

Hershkovitz et al. (2008) examined samples of bone from a woman and infant buried together at Atlit-Yam, a now-submerged Neolithic site in the eastern Mediterranean just off the coast of northern Israel. The samples date to about 8,000–9,000 years ago. The researchers used polymerase chain reaction to find DNA sequences from *Mycobacterium tuberculosis* and high-performance liquid chromatography to identify lipids containing mycolic acid, a marker for the presence of *M. tuberculosis*.

Cancer

Cancer may be a curse of the modern world, but it is not new. Examples have been found in the skeletons of early humans, and of course the disease is common among many other species, though it seems to occur only in vertebrates. In mammals, it is rare in wild animals and much more common in domesticated animals (Capasso 2004). A few cases have been found in dinosaur fossils.

Finding cancers in ancient fossils or other human remains is complicated. Cancer was probably less frequent in ancient populations because life spans were shorter then—prehistoric humans tended to die before a cancer could develop. Our modern lives are much longer, but also less healthy in some ways. Many cancers today result from environmental factors (e.g., smoking, air pollution) and a lifestyle that did not exist in ancient times, and those types of cancers are unlikely to be found in fossils. Other types of cancer that occur in blood or soft tissues won't be visible in skeletons. Furthermore, a number of factors—soil acidity, breaks and abrasions after death, actions by microorganisms—can produce changes in skeletal remains that may look like the effects of cancers. Still, some cancers (such as osteosarcoma) are independent of those factors and thus probably occurred as frequently in early humans as they do today.

Evidence of cancers is rare in human fossils, but some tumors have been found. The oldest so far is a case of osteosarcoma in South Africa 1.7 million years ago. Using advanced three-dimensional imaging techniques, Odes et al. (2016) found the tumor in a hominid metatarsal specimen. Monge et al. (2013) found an early example in a 120,000-year-old Neanderthal skeleton in Croatia. A rib showed good evidence of a primary osteolytic lesion. A large number of Egyptian mummies, where soft tissue has been preserved to some extent, have been examined by various methods, but few cancers have been found. Using X-rays and multidetector computerized tomography, Prates et al. (2011) found evidence of focal dense bone lesions in the spine, pelvis, and proximal extremities in a mummy of Ptolemaic Egypt. The most likely diagnosis is osteoblastic metastatic disease that originated in the prostate. Klaus (2018) reported four possible cases of prostate cancer in Peru from about 900 to 1600 CE. He found abnormal bone formation in the lumbar vertebrae and used differential diagnosis to eliminate other potential causes (such as infectious diseases and bone disorders). However, metastatic prostate cancer could not be eliminated, and the findings were consistent with this diagnosis. Multiple (mainly osteolytic) lesions were found in the remains of a young male from Amara West in northern Sudan from 1200 BCE (Binder et al. 2014). The lesions were on the vertebrae, ribs, sternum, clavicles, scapulae, pelvis, and humeral and femoral heads. This might represent the first finding of a metastatic cancer. Schultz et al. (2007) examined the remains of a forty-to-fifty-year-old Scythian king from southern Siberia dating back to the Iron Age, 2,700 years ago. They found osteoblastic and osteoclastic lesions (two types of bone metastases) throughout his entire skeleton.

Prehistoric Warfare

Wars and conflicts have been a part of the human experience for all of recorded history. Were they also common in prehistoric populations, and did those conflicts have any influence on evolution? Some anthropologists have argued that prehistoric humans rarely engaged in conflicts because population densities were so low. Others have speculated that the effort to avoid overlap of hunting areas and the resulting potential for conflict was a factor in the migration of early hominids out of Africa. However, still others believe aggression was common and even rewarded evolutionarily (McDonald, Navarrete, and Van Vugt 2012). The actual evidence of prehistoric warfare is somewhat limited, but in an extensive review, Gat (2015) documented a significant number of cases of conflict among early humans. Bowles (2009) modeled the effects of early warfare on human populations and found that it had a substantial effect on human evolution.

Cranial trauma can sometimes be an indicator of violence. Jiménez-Brobeil and Oumaoui (2009) looked for this type of evidence in the remains of 410 prehistoric individuals from the Neolithic to the Bronze Age on the southeastern Iberian Peninsula and compared them to 267 medieval and modern skulls. They also studied 267 crania from medieval and modern times for comparative purposes. Injuries were more common in adults than in juveniles and also more common in males than in females. The injuries were consistent with interpersonal or intergroup conflict.

Interestingly, humans are not the only species to engage in some type of warfare. Our closest relatives, the chimpanzees, also engage in intergroup aggression, mostly as raids and ambushes. The reason is not clear, but it has been speculated that it has to do with expanding their territory, access to food supplies, and males' access to females (Wilson and Wrangham 2003). A study of two groups of chimpanzees in Uganda (Mitani, Watts, and Amsler 2010) found that one group greatly expanded its territory, resulting in the deaths of many of the other group. These observations support their hypothesis that the chimpanzees use aggression to seize more territory.

Did Lucy Fall to Her Death?

In 1974, Donald Johanson found one of the most complete and most famous hominid skeletons in the Afar region of Ethiopia (Johanson et al. 1982). They named her Lucy. She lived about 3.2 million years ago. Her group, *Australopithecus afarensis,* is thought to be either a direct ancestor of modern humans or closely related to such a direct ancestor. The skeleton showed that Lucy was about the size of a chimpanzee—she was three feet six inches tall and weighed a little over sixty pounds. Her hip and leg bones were similar to those of modern humans and showed that she walked upright.

Kappelman et al. (2016) examined the skeleton to try to find evidence for how Lucy died. They noted that the condition of the bones is consistent with multiple compressive and hinge (greenstick) fractures. They conclude that Lucy fell from a tree and suffered multiple fractures and probably injuries to internal organs, and that these injuries caused her death.

Not everyone agrees with this assessment, as fascinating as it is. First, Johanson believes that the damage seen in the bones is consistent with normal weathering and aging. Studies of the skeletal remains of humans can be challenging. They are often affected by many factors, such as fungi, bacteria, and other microorganisms, plant roots, insects, worms, and water. The state of preservation is also a major factor. Moreover, how these agents affect remains is incompletely understood, and so attributing changes in bone morphology to diseases can be difficult. Finally, the study makes a lot of assumptions. Behrents (2016) discounts those because we have no way of knowing the circumstances of Lucy's death. All we know is that she died. Anything else is mere supposition.

Shift to Farming

The change from hunter-gather societies to small farming communities about 6,000 years ago was a major milestone in human development. Food supplies became more stable, and people could begin to specialize in various occupations. However, injuries and infectious disease continued to be major causes of death (noncommunicable diseases, such as cancer, obesity, diabetes, and hypertension, were rare).

At about the same time as humans began farming, a number of human diseases seem to have increased. These include smallpox, measles, malaria, pertussis, schistosomiasis, and tuberculosis. While these diseases existed before, some scientists speculate that their emergence was facilitated by increased populations and population densities that could sustain the pathogens. Also, the presence of domesticated animals might have provided an alternative host for some diseases (Pearce-Duvet 2006). Finally, as humans began to farm, they radically changed the environment, and those changes might have contributed to the rise of diseases. For example, farming-related changes might have increased the interactions between humans and disease vectors.

Death in the More Recent Past

In the seventeenth and eighteenth centuries, conditions in European cities were deplorable. In London, the water and sewage systems were primitive, resulting in much disease and many deaths. Landers (1987) found that London had high rates of death that could be attributed to the population density of the city. However, deaths varied by season. Landers and Mouzas (1988) found that in the seventeenth century, mortality was higher in the summer, likely from gastric infections. In the eighteenth century, deaths peaked in the winter months and likely represented respiratory infections and typhus. Cummins, Kelly, and Grada (2015) used data from 930,000 burials and 630,000 baptisms to study patterns of birth and death in London from 1560 to 1665. In those years, the plague regularly afflicted the city and killed many, but the plagues of 1563, 1603, 1625, and 1665 killed more people than usual. Interestingly, the rich fared better than the poor. In fact, these epidemics did not begin at the port, as previously believed. They began in the poor sections of London.

For centuries, bacterial infections killed many people. A simple cut could fester into gangrene. Appendicitis and peritonitis were often fatal. In 1928, that began to change. Alexander Fleming noticed that spores of a green mold (*Penicillium chrysogenum*) prevented the growth of bacteria. He suggested that the mold secreted a substance that killed the bacteria. Later others isolated the first antibiotic, penicillin, from this mold. In the years following this discovery, additional antibiotics were isolated from fungi; more recent versions were developed by chemical synthesis. Those discoveries provided outstanding medicines to treat infectious diseases.

The *New England Journal of Medicine* published a survey of medical papers from 1812 and 1912 to show one measure of the differences in medical cases in those eras (Jones, Podolsky, and Greene 2012). As one might guess, there are differences and similarities among the three lists. The 1812 list of cases includes familiar problems, such as gunshot wounds, spina bifida, tetralogy of Fallot, diabetes, hernia, epilepsy, osteomyelitis, syphilis, cancer, croup, asthma, rabies, urethral stones, angina, diarrhea, and burns. Others are less familiar to us today, including apoplexy (an old term for a cerebrovascular accident or stroke), wounds from cannonballs, spontaneous combustion, and a multitude of fevers. The 1912 reports focused more on infections, such as tuberculosis, gonorrhea, syphilis, diphtheria, measles, pneumonia, scarlet fever, and typhoid. The list of cases also included diseases familiar to us today, such as cancer, eclampsia, impotence, arthritis, and heart disease (infectious or valvular, but not atherosclerotic). The 1912 papers optimistically predicted that preventable diseases would be eliminated and cancer cured by 1993, but they also included warnings of the perils of modern life, including "automobile knee" and the hazards of inactivity.

Pandemics and Other Large-Scale Events

While each death is of an individual, there have been a number of times when very large numbers of deaths occur within a short time frame. Most of these events have been pandemics, but at least one war resulted in equally huge numbers of deaths.

The Antonine Plagues

This pandemic, which is now believed to be an outbreak of smallpox that killed 5 million people, is named for the emperor Marcus Aurelius Antoninus, and occurred between 165 and 180 CE. The emperor was one of its victims.

Cholera

Cholera likely originated in India along the Ganges Delta centuries ago. It is caused by a bacterium, *Vibrio cholerae*, that is transmitted in impure water or in contaminated food. Different strains result in different degrees of illness. It is primarily a disease of poor countries, rarely occurring in developed nations. According to the WHO, each year there are 3 million to 5 million new cases, and 120,000 people die.

The Black Death or Bubonic Plague

Bubonic plague is caused by a bacterium called *Yersinia pestis*. The name Black Death comes from the black spots that form on the skin of those infected with the disease. It is thought to have originated in Central Asia and to have been transported to Europe in fleas on black rats carried on trading ships.

In 541 CE, rats on Egyptian grain boats brought a plague to the Eastern Roman Empire. It even infected Emperor Justinian I (after whom this wave of disease is called the Justinian plague). He survived, but 25 million people died. This is thought to be the first recorded outbreak of the bubonic plague.

While plague was fairly common in medieval and early modern times, the Black Death that swept through Europe in the middle of the fourteenth century, peaking in 1346–1353, killed between 75 million and 200 million of the estimated 450 million people on Earth at that time.

Death of Native Americans

Native Americans arrived in the Western Hemisphere 15,000–20,000 years ago and eventually inhabited much of North and South America. Population estimates vary greatly, but just before the arrival of the Europeans in the very late fifteenth century, Native Americans numbered somewhere between 8 million and 102 million (Nunn and Qian 2010). Until the arrival of Europeans at the end of the fifteenth century, they were isolated from other gene pools; they had adapted to their new environment and developed immunity to the diseases that were here.

That all changed with the arrival of the Europeans. In addition to new weapons and horses, the Europeans brought diseases that the Native Americans had never experienced, such as smallpox, measles, whooping cough, chicken pox, bubonic plague, typhus, and malaria. Completely lacking immunity, the Native Americans were defenseless. The spread of diseases was not always unintentional: colonists sometimes deliberately provided Native Americans with blankets and other items from sick Europeans. Their population plummeted. As one example, the population of Mexico before the arrival of the Europeans was between 25 million and 30 million. Within forty years after Cortés conquered Mexico in 1519, it was 3 million. Other groups suffered similar losses.

World War II

World War II resulted in the deaths of between 70 million and 85 million people. This includes combatants and noncombatants in both the Pacific and European theaters. Of these, a significant number of deaths (19 million to 28 million) were caused by disease and starvation and thus were indirect results of the combat.

1918 Spanish Flu

For most of us, influenza is a nuisance that we have to deal with every year or so, especially if we forget to be vaccinated that year. However, the Spanish flu of 1918 was much more than that. In fact, it might be the most lethal pandemic of all time. That year was the final year of World War 1, which had itself cost many lives. However, the Spanish flu that year killed 20 million to 40 million people, including 675,000 Americans, far more than were killed in all of the war. Although this pandemic is called the "Spanish" flu, it did not originate in Spain. No one knows exactly where it began, but most believe it came from China. The genetic sequence of the Spanish flu has been analyzed by scientists in an effort to determine what made it so virulent.

HIV/AIDS

The HIV/AIDS pandemic seems to have originated in Africa, in Cameroon and the Congo. The first cases were identified in the early 1980s, but later studies have found blood samples from much earlier (at least as early as the 1950s) that contain the virus. The introduction of highly active antiretroviral therapy has changed HIV infection from a death sentence to a chronic infection for those with access to the drugs. Estimates from 2012 cite over 35 million cases of HIV infection worldwide.

Looking to the Future: Diseases of Lifestyle

We are now entering a new phase that is even less certain. Some scholars estimate that we are near to a time when life expectancy will surpass 100 years. Medicine continues to advance. Our understanding of disease and biology improves every year. On the other hand, a growing worldwide epidemic of obesity, diabetes, metabolic syndrome, and heart disease is under way. For the first time in US history, the life expectancy of the next generation might be less than that of the previous one.

The root of this problem seems to be the widespread availability of food. The world is awash in calories. For the last fifty years, most people in the developed nations have had easy access to food, and unfortunately, much of that food has contained high numbers of calories and lots of salt and sugar. "Supersize" has become the typical portion. The result is an increase in what are called lifestyle diseases (e.g., obesity, heart disease, diabetes, high blood pressure, metabolic syndrome). These diseases are the major causes of death in the developed world. Today even many developing nations have plenty of food, and increasingly, they are beginning to see the same disease patterns as the United States and the rest of the developed world.

One possible explanation for the increase in incidence of these diseases is called the "thrifty genotype" hypothesis (Neal 1962). The idea is that for millennia, humans lived in environments where food was scarce. Our ancestors' bodies adapted by favoring genes that enhanced the body's ability to store fat (insulin resistance). Every calorie counted, and so our bodies evolved to make the most of the food that we ate. In our hunter-gather ancestors, the thrifty genes allowed people to fatten more quickly in time of plenty. This was especially important for childbearing women so that they could remain healthy enough to endure the hardships of pregnancy. This strategy worked in an environment that provided limited food and required maximum expenditure of energy. Today our situation is quite different. We have plenty of food, and many of us engage in little physical activity. We have become a nation of couch potatoes. However, the change in our environment occurred very fast in evolutionary terms, and our genes have not caught up. Thus, we are still trying to extract every bit of energy from an abundance. The result is an accumulation of energy stored in the form of fat. Obesity, high blood pressure, diabetes, metabolic syndrome, and early death are increasing.

The thrifty gene hypothesis has its problems. Many of the populations that have experienced weight gain in times of plenty have no evolutionary history of famine. For example, Pacific Islanders existed in a paradise of readily available food for very long historical periods. Also, modern hunter-gathers do not become obese in times of plenty.

Alternatives have been suggested that involve the environment experienced by the developing fetus and its relationship to insulin resistance and obesity in later life. Other scientists have suggested that epigenetics (changes in gene function that do not involve changes in DNA sequence), environmental factors, specific gene mutations, and other factors might be involved. However, the question remains unanswered. Clearly, Neal's original hypothesis set in motion an intriguing line of experimentation. No matter how the increase in obesity, diabetes, and metabolic syndrome came about, those diseases are a serious threat to the health and life span of Americans and others.

Neurodegenerative diseases are an increasing public health problem. In the United States today, 5.8 million people are living with dementia caused by Alzheimer's disease (Alzheimer's Association 2019). The toll on their families and caregivers is enormous. The current annual cost of this disease is $290 billion. There are no cures and no treatments that change the course of the disease. By 2050, the number of Americans with Alzheimer's disease is expected to rise to 14 million, and the cost will rise to $1.1 trillion. Other neurodegenerative diseases (e.g., Parkinson's disease, amyotrophic lateral sclerosis, frontotemporal dementia, and multiple sclerosis) are expected to increase also as our population ages. Many other nations will have similar challenges.

Another phenomenon has gained a lot of attention in recent years. Investigators have noted a marked increase in mortality for middle-aged white (non-Hispanic) men and women between 1999 and 2013 in the United States (Case and Deaton 2015). No other country saw such an increase, and the life span of every other group in the United States continued to increase. Deaths from drug and alcohol abuse, suicide, and cirrhosis are mainly responsible for the decrease among this particular group. Furthermore, the decrease in life span correlated with those who had less formal education. Those years also saw economic hardships for many blue-collar workers.

Most disheartening is that the current generation of young Americans might be the first to actually have a shorter life span than their parents (Olshansky et al. 2005). Since about 1850, people have been living longer. The life expectancy at birth has been increased by three months for every year since then. Now, however, that rate of increase is at risk. Olshansky et al. (2005) state, "We see a threatening storm—obesity—that will, if unchecked, have a negative effect on life expectancy." After being relatively stable for some time, obesity rates increased by 50 percent per decade during the 1980s and 1990s. The lifetime risk of diabetes has also increased dramatically. Obesity significantly shortens life expectancy. Estimates suggest that life expectancy at birth could increase by about 1/3 to 1 year, but that obesity will shorten lives by more than all accidental deaths combined.

4

Aging and Senescence

Old age is not a battle. Old age is a massacre.

—Philip Roth, *Everyman*

Roth's "massacre" is old age, and it is familiar to most of us. We all age, and we all die, but death is not necessarily linked to aging. Even infants die. Clearly, though, aging affects many of our organs and tissues, our ability to function, our quality of life, and our ability to fend off diseases and conditions that do lead to death. What is aging, after all? Do all living things age? Why do people age? Do individual people age differently? Can we do anything about it? These are fascinating questions, and they almost surely have their answers somewhere in genetics and evolution (Charlesworth 2000).

What Are Aging and Senescence?

The terms *aging* and *senescence* have considerable overlap in meaning. *Senescence* usually refers to the loss of biological functions in cells or whole organisms. Senescent cells lose their ability to divide and begin to show changes to chromatin, their secretome, and activation of tumor suppressors (van Deursen 2014). Chromatin is the chromosomes as they exist in a less-condensed state except for when they are dividing. The secretome or senescence-associated secretory phenotype (SASP) includes many different compounds, including cytokines, chemokines, growth factors, and proteases. SASP has been implicated in chronic inflammation, tumorigenesis (the origins of tumors), and poor stem cell renewal (cells that can develop into any type of cell in the body) (Basisty et al. 2020) as well as a number of aging-related conditions, such as atherosclerosis, osteoarthritis, cancer, and more.

We all have some idea of aging. We see its effects in people all around us and in ourselves: the laugh lines, the salt-and-pepper hair, the receding hairline, the aches and pains. We see them first in our grandparents and parents; later on, we begin to see them in our friends and ourselves.

The Biology of Death. Gary C. Howard, Oxford University Press. © Oxford University Press 2021.
DOI: 10.1093/oso/9780190687724.003.0004

And we often don't like what we see. In *The Picture of Dorian Gray* by Oscar Wilde, Gray was willing to sell his soul so that he could stay young. Very few Americans would be willing to sell their souls to stay young, but many are happy to spend lots of money to try to slow the aging process, even if just a bit. For years, companies have advertised various concoctions to remove the wrinkles from skin, put color back into hair, regrow hair, increase stamina, and improve muscle tone. The "cures" often capitalize on scientific buzzwords to sell their products. Recently, stem cells have become the rage, and therapies based on them have popped up like mushrooms. Turner and Knoepfler (2016) estimate that throughout the United States there are 351 companies operating out of 570 clinics peddling therapies based on various types of stem cells, including nonhomologous, autologous, allogeneic, induced pluripotent, and xenogeneic. Many are promoted as treatments for different types of aging, but none of these therapies is approved by the Food and Drug Administration. More recently, products based on the genome-editing technology CRISPR have begun to appear.

Humans and many other animals and plants change throughout their lives. During the development of a fertilized egg and on through infancy, the teen years, and early adulthood, those changes are generally positive. We grow in size and strength. Vision becomes more accurate. We gain in our mental powers. After this, much of adulthood seems to be a period of stasis. Over time, inevitably Roth's "massacre" begins, and so does the inexorable march to death.

Physiological Manifestations

Anyone over the age of fifty certainly will recognize the physiological changes that accompany aging. While some of the changes may threaten one's vanity, they are not significant health threats. Others are much more serious.

Hair

Hair begins to lose its color and in some cases it starts to thin out, especially among men. Hair color is due to melanocytes, pigment cells in the hair follicles that produce melanin. With age, the pigment cells begin to die off, and less pigment is produced. Graying can occur at any age, and that age is determined mostly by our genes. Smoking and a number of health conditions can cause hair to go prematurely gray, though stress does not. However, interestingly, our immune system may be involved in the loss of hair color (Harris et al. 2018). Immune cells secrete interferons, which are signaling molecules that induce an immune response. In mice, melanogenesis-associated transcription factor represses expression of innate immune genes in cells in the melanocyte lineage. When laboratory mice were induced to activate their innate immune response, their hair turned gray. These results suggest a connection between the immune system and hair color.

Losing hair is normal. In fact, most of us lose about 100 hairs every day. Those hairs are replaced by new hair growth, but over time, the balance shifts so that we lose more hairs than we produce. Hair loss is caused by a number of factors, but the most common is our genes. Men typically begin to notice a receding hairline and bald spots on the back of the head (male pattern baldness). Women who experience hair loss usually notice that their part is getting wider (female pattern baldness). There are many reasons for hair loss besides aging, though, including hormonal changes, medications, tight hairstyles, stress, and radiation or chemotherapy.

Skin

The skin is the largest organ in our body and serves as the interface of the body with the environment. Aging skin is one of the most noticeable characteristics of aging, and lots of money is spent trying to prevent it. The aging of skin occurs in two forms (Tobin 2017). Intrinsic aging is governed mostly by genetics. As the skin ages, the rete ridges, or pegs that anchor the epidermis to the underlying tissue, erode. However, intrinsic aging affects one's external appearance only slightly: the skin is drier and less elastic, with fine wrinkles. Extrinsic aging is caused by external factors, and the two most significant of these are ultraviolet light and smoking. The ultraviolet radiation of sunlight causes the skin's collagen and elastin fibers to break down; without that important connective tissue, the skin loses its structure and flexibility, and the results are coarse wrinkles, creases, and sagging skin (a condition called solar elastosis). Another aging-related condition is telangiectasia, when tiny blood vessels called venules become wider and result in threadlike red lines on the skin, sometimes called spider veins. Elastosis increases with age in men and women, but telangiectasia tends to be more common in men. Smoking increases the risk of both conditions (Kennedy et al. 2003) and is an important risk factor for developing cutaneous squamous cell carcinomas.

Frailty

One of the adjectives often applied to the elderly is *frail*. In this context, frailty refers to a lack of reserves to respond to stresses. But unlike baldness or sagging skin, frailty is a very serious condition. Its manifestations include weight loss, poor hand grip strength, slow gait, exhaustion, and reduced physical activity. Those who are frail fall more often, move less, are hospitalized more, and die earlier than more robust people. Frailty often goes hand in hand with the loss of bone, muscle mass, and balance. It is sometimes reversible with treatment of the underlying pathologies and improved nutrition and exercise. Although the causes of frailty are unknown, several risk factors have been identified, including inactivity, loss of appetite (anorexia of aging), loss of nutrients, loneliness and other social factors, and endocrine changes. Multimorbidity, or the presence of multiple disorders, characterizes many of those who are frail. Also involved are catabolic diseases that result in the loss of muscle and bone mass.

Unfortunately, the risks associated with frailty are exacerbated by socioeconomic status. Those with few resources are at increased risk of becoming frail (Kehler 2019). Frailty is on the rise as populations around the world are aging. It is becoming a major public health crisis. Sadly, the concept continues to evolve, and no standard instrument is in place for clinicians to identify frailty (Hoogendijk et al. 2019). The cost of care for frail individuals is also climbing.

Senses

The senses are affected by aging, and their loss can have serious effects. First, vision deteriorates. Focusing on close objects becomes more difficult. Older people can no longer distinguish small changes in spatial contrast. The number of light-sensing cells in the retina drops. Less light reaches the retina in older eyes. The pupil tends to constrict more, and the lens becomes denser (Owsley 2011). Cataracts, age-related macular degeneration, and glaucoma are also more common in older patients. Older people often complain of not being able to see in dim light; many do not like to drive at night. There is good reason for this. By age sixty to seventy, the number of rods in the eye is reduced dramatically, and as a result the eye does not adapt to the dark as quickly. In addition, the visual pigment rhodopsin, which helps enable vision in low light, is not regenerated as quickly in older eyes.

Second, older people are often hard of hearing. Hearing loss often is associated with other issues, such as cognitive decline, depression, poor balance and falls, and greater risk of dementia and death (Davis et al. 2016). In addition, it affects communication, social isolation, autonomy, driving, and finances. Since hearing loss is so gradual, it can be unnoticed for some time. The ability to hear high-pitched sounds is usually the first to be lost. Later, it becomes more difficult to carry on a conversation in a crowded room, and in general, understanding speech can be challenging. There are no cures, but hearing aids, cochlear implants, and other devices can help.

Third, both taste and smell are affected by aging (Boyce and Shone 2006). Aging results in fewer fibers in the olfactory bulb and fewer olfactory receptors. Sensory cells in the mucosa of the nose and mouth are lost, and the decline in the sense of taste may be related to the loss of the ability to smell food. Tooth loss and the use of dentures can reduce the ability to chew, which might also contribute to a loss of taste. However, some of the effects of aging on the senses may be due to a more general decline in cognitive ability.

Smell and taste are not easy to measure, and there are many possible causes of a loss of smell or taste. Loss of these senses is often ignored as trivial and not important to health. However, without them, seniors might eat less and thus lose weight and strength. Finally, to compensate for the loss of taste, seniors tend to add more sugar and salt to their meals (Stevens et al. 1991), and this action can indirectly affect their health and wellness.

Heart

Heart disease is the major killer in the United States and the developed world, and the major risk factor for heart disease is age. More than 70 percent of Americans aged sixty to seventy-nine have heart disease, and more than 80 percent of those over eighty have it (Go et al. 2014). Even without disease, aging hearts have several features that are independent of the normal risk factors, such as smoking, high blood cholesterol levels, and high blood pressure. In older people, the left ventricle of the heart is more likely to be enlarged (hypertrophy). The heart has less ability to fill with blood (decreased diastolic function), and while it may still be able to pump blood well (systolic function) at rest, it does not function as well when exercising (Chiao and Rabinovitch 2015). There is also an increased risk of unsynchronized contractions of the atrial muscle (atrial fibrillation).

The causes of the changes to aging hearts are unknown, but several possibilities are being examined. For example, two biological pathways (a pathway is a series of interactions among molecules that affect cell function) have been implicated in these declines: mechanistic target of rapamycin (mTOR) and insulin-like growth factor (IGF-1). mTOR is a master regulator that is involved in many cellular functions, such as sensing nutrients, maintaining proteostasis, autophagy, mitochondrial dysfunction, cellular senescence, and reduced stem cell function. Another example involves cardiomyocytes, the beating cells in the heart. These cells are post-mitotic (that is, they can no longer divide to replace themselves) and so damage to them in aging hearts is especially critical, as the heart no longer has the ability to replace damaged cells. The production of reactive oxygen species (ROS, highly reactive molecules that can damage cells) is enhanced in aged hearts, and the cells' ability to neutralize ROS is reduced (Dai et al. 2014). IGF-1 is involved in critical cellular functions as well, such as cell growth, differentiation, development, and tissue repair, and disruptions of any of these could hinder heart function.

Changes to the extracellular matrix (ECM) may also be involved. Produced by cardiac fibroblasts, the ECM includes several proteins, such as collagen, fibronectin, fibrinogen, elastin, and laminin, and provides structure and support to the cardiomyocytes; in essence, it forms a very complex scaffold that cells can adhere to. The production of these proteins or structural reorganization within the ECM can get out of balance, resulting in heart muscle that is too stiff to function efficiently.

To pump blood, the heart must contract and relax, and that muscle action requires a cycling of calcium ions that affects the heart muscle's actin and myosin fibers. If that cycle is not functioning correctly, the heart will not relax effectively. Heart enlargement and hypertension are closely related to angiotensin II. A number of microRNAs—short RNA sequences that regulate proteins in cells—have been linked to aging hearts.

Bones and Muscle

Muscles lose their strength as people age. As a result, many people have trouble with ordinary daily activities. The decrease in lean muscle mass that accompanies aging reduces the cross-sectional area of muscle and seems to be more a result of a loss of muscle fibers than a reduction in size of individual fibers (Williams, Higgins, and Lewek 2002). Older muscle also contains less contractile tissue and more noncontractile tissue than younger muscle, so strength drops. Fat is deposited in muscle, and in some areas (such as the hands), tough fibrous tissue replaces muscle. Those changes can affect the gait and posture and can slow movement. Training can improve muscle endurance and strength, and it can also potentially prevent muscle loss in the first place.

Bones provide support and structure to the body. As we age, the bones lose mass (calcium and other minerals) and become more brittle (Boskey and Coleman 2010). The cartilage that cushions the bones in the joints becomes worn or damaged. The discs between the vertebrae in our spines lose fluid over time and become thin. We might lose height because of this compression of the vertebrae, and also because of the increased curvature of the spine. Deposits of minerals may form in some joints, such as the shoulder. Damage to the joints may result in inflammation and stiffness and eventually to arthritis.

Women are at particular risk for the serious loss of bone density that is called osteoporosis, accounting for about 68 percent of cases. Part of the reason is that they tend to have thinner bones than men, and at menopause, the loss of estrogen accelerates their bone loss. Half of all women over age fifty will likely have a bone fracture that is related to osteoporosis.

Digestion

The digestive system (esophagus, stomach, small intestines, colon, rectum) is affected by age (Salles 2007). Digestion slows. Peristalsis, the contractions of the esophagus and intestines that move food along through our system, becomes weaker. Gastroesophageal reflux, the leakage of stomach acid back into the esophagus, is more common in older individuals. The amount of acid produced by the stomach seems to be unaffected, but it is not clear if the stomach is slower to empty. However, transit time in the colon is increased. Disruptions of the digestive tract can be serious for older patients. They often suffer from a decreased interest in food because of a loss of taste and smell, dental problems, and depression. Reduced intake of food can exacerbate other conditions and deprive them of needed nutrition.

Bladder

As the bladder ages, its capacity decreases, and uninhibited contractions increase, which can result in the need to urinate often and urinary incontinence (Siroky 2004). Urinary flow rate is reduced, and more urine is retained in the bladder after voiding. However, the mechanisms underlying these changes are not clear, and they are often confounded by disease conditions, such as prostate enlargement. Nevertheless, a few mechanisms have been implicated. The detrusor muscle is the smooth muscle that lines the bladder. It is normally relaxed, but contracts to expel urine. It seems to function less effectively in older people. The bladder wall becomes more fibrous with age, and that might interfere with muscle contraction. The bladder becomes more sensitive to the levels of neurotransmitters (e.g., norepinephrine).

Mental and Cognitive Issues

The brain changes as we age (Harada et al. 2013). Some of the changes are physical, and others involve cognitive abilities. Physical changes to the brain begin in a person's sixties and include changes to the frontal lobe and hippocampus, areas of the brain that are involved in cognitive function. About 60 percent of cognitive decline is estimated to result from genetics (McClearn et al. 1997).

Though people often speak about "shrinkage," aging brains do not actually shrink in the sense of decreasing significantly in volume. The number of neurons remains about the same, but the neurons get smaller and have fewer connections to other neurons (Resnick et al. 2003). In addition, neurons experience other changes, including shorter dendrites, fewer neuritic spines, and a general reduction in the complexity of the dendritic arborization (Dickstein et al. 2007). Neurons need multiple connections through their dendrites to other neurons to remain healthy. The cortex thins as the number of connections is reduced. The thickness of the myelin sheath that serves as an insulator along nerve fibers declines, so processing is slowed. Levels of neurotransmitters, such as dopamine, acetylcholine, serotonin, and norepinephrine, are reduced, and that reduction might hinder cognition and increase the tendency to depression.

Although the relationships are not well understood, the physical changes in the brain might contribute to a decrease in cognitive function. A certain amount of decline is inevitable, but the changes vary from person to person. For example, as we age, it becomes more difficult to memorize new information. Multitasking is more difficult too, and we accomplish tasks more slowly. We have a harder time remembering names and appointments. Some older individuals retain their mental faculties well into their nineties. Other struggle and even slip into dementia. However, Alzheimer's disease is not part of normal aging.

Cognition is often divided into two types. Crystallized intelligence includes those things that are learned and practiced, such as vocabulary and general knowledge. It is stable with aging and might even improve somewhat. Fluid intelligence is independent of what a person has learned. It involves processing speed, attention, memory, executive functioning, and the ability to manipulate new information. It peaks in the third decade and declines slowly. Older people sometimes compensate for the loss of fluid intelligence with enhanced crystallized intelligence.

Processing speed is just what it sounds like: the speed at which an individual processes information. As we age, this ability declines, and we take longer to think things through. Attention refers to the ability to focus, but is more than just that. It also refers to the ability to pay attention in the face of distractions, such as conversing in a noisy environment or doing two things at once. Memory is divided into two types: declarative (the recollection of facts and events) and nondeclarative memory is the recollection of how to do simple tasks, such as tie one's shoes or ride a bicycle. Language is the use of vocabulary, and this remains strong throughout life. Executive functioning is the ability to control our own lives by planning, organizing, and problem-solving.

Dementia is perhaps the most insidious disorder of aging. It robs us of our language, ability to think, memories, and personality. It destroys the very things that define us as individuals. Dementia is a major manifestation of Alzheimer's disease, Parkinson's disease, and frontotemporal dementia. The causes of these diseases are unknown, but each features aggregations of a specific protein. In Alzheimer's disease, it is plaques of the amyloid beta peptide and neurofibrillary tangles of the protein tau. In Parkinson's disease, the aggregations are of alpha-synuclein. In frontotemporal dementia, it is TDP-43. How the aggregations are related to neuronal death and dementia is not clear. Tragically, all of these diseases are progressive and ultimately fatal, and no therapies are available.

Does Everything Age?

Aging or senescence seems to be inevitable for living organisms, but is that true for all animals and even plants? Animals age at different rates. Most theories of aging predict an increase in mortality and a decrease in fertility as an organism ages. But is that really true for all species? Do some animals or plants live essentially forever? Is there any pattern to aging?

The answers are mixed. One study compared forty-six species (eleven mammals, twelve other vertebrates, ten invertebrates, twelve vascular plants, and one green alga) for patterns of mortality and fertility (Jones et al. 2014). For some, fertility increases as they age. For others, both fertility and mortality spike in old age. Still others are fertile throughout their lives. Animals show a lot of similarity, and plants are much more diverse. A big problem for the study was how to organize the data so that comparisons could be made. The researchers chose to define "relative mortality" and "relative fertility" numbers for each species. They calculated these numbers by dividing mortality or fertility at a specific age by the average mortality or fertility across the organism's whole lifetime.

Plants can live for a very long time. Unlike animals that have a defined set of cells to produce sperm and eggs, plants do not. They produce seeds at multiple sites through the plant, and each of those sites generates cells that produce pollen and eggs. Somatic mutations (mutations in nonreproductive cells) can be carried into what will become reproductive cells so that those mutations can enter the germ line. Part of the reason for their longevity may be that their stem cells, which are located in the meristems (growing parts) of the plant, do not divide very often (Burian, de Reuille, and Kuhlemeier 2016), even as they provide lots of precursor cells for the different parts of the plant. The relatively few divisions suggest that the stem cells do not age or senesce, and thus the plant can continue growing for a very long time. Also, the plant's various meristems seem to operate independently of one another, which means that a mutation in one would only affect that specific branch; the rest of the branches would be safe. A study of a 238-year-old oak tree in France on the campus of the University of Lausanne called the Napoleon Oak supports these ideas (Schmid-Siegert et al. 2017). By comparing DNA sequences from various parts of the tree, they showed that the stem cells were protected from mutations.

Plants do have an immune system, although it is not as robust as the human immune system. Wang, Cui et al. (2020) examined the immune genes in the vascular cambium of ginkgo trees (*Ginkgo biloba*), which also have remarkably long lives—some ginkgos have lived more than 1000 years. They found that the genes do not lose their activity with age, unlike those in humans. This ageless immune system gives the ginkgos a greater ability to fend off diseases and to survive longer.

Amazingly, some organisms do not seem to age at all. When food is short or it is injured, the jellyfish *Turritopsis dohrnii* has an unusual ability to sort of re-create itself. Normally, the jellyfish begins as a fertilized egg that grows into a larva. The larva attached itself to a surface and develops a tube-shaped structure called a polyp. The polyp will later bud off a segment that will develop into a small jellyfish-like organism called an ephyra that grows into a full-sized medusa. When faced with a life-threatening stress, the medusa reverts to a blob of tissue that transforms back into a sexually immature polyp stage. So far, *Turritopsis dohrnii* is the only jellyfish with this ability.

Jellyfish are not the only organisms that seem to defy aging. For example, the genus *Hydra* includes freshwater relatives of corals, sea anemones, sea urchins, and jellyfish. They do not seem to age (Tomczyk et al. 2015) and have become a valuable model for aging studies. Organisms in this genus have stem cells that are able to repair damage that occurs over time. The FoxO genes regulate stem cell life span in many organisms, and they are very active in *Hydra*.

Naked mole rats (*Heterocephalus glaber*) also have much longer life spans than expected. They live up to thirty-two years, the longest of any rodent (compared to about four years for a house mouse, *Mus musculus*). Their mortality rate does not increase with age, and they are very resistant to cancer. The reasons for these abilities are unknown, but several ideas have been suggested. First, they can reduce their metabolism and avoid oxidative stress when food is short. Second, they express high levels of DNA repair genes. Third, the animals secrete a hyaluronan (a type of negatively charged polysaccharide or polymer of sugars that is five times larger than that seen in humans or mice (Tian et al. 2013). Hyaluronan is an important component of the extracellular matrix (the scaffold surrounding cells) and contributes to proliferation of cells during growth and their migration to an appropriate position in the developing organism. The larger hyaluronan molecules might give the mole rats' skin the elasticity needed to live in tunnels, where the skin is constantly exposed to rough surfaces, and that trait might also provide resistance to cancer and a longer life span. The naked mole rats have much more of this compound than other organisms because they have decreased activity of enzymes that degrade it and a unique sequence in the hyaluronan synthase 2 (*HAS2*) gene. If the amount of hyaluronan is lowered by genetic means, the rats become much more susceptible to cancer.

Interestingly, eukaryotic cells seem to have a limited life span in culture (Hayflick and Moorhead (1961). This upper bound has become known as the Hayflick limit, and it means that a normal human cell cannot divide indefinitely, but only forty to sixty times. Once the cell reaches that limit, it dies by apoptosis—that is, the program for cell death kicks in. The Hayflick limit holds for both embryonic and adult cells. Researchers later described a cellular mechanism that limits the viability of cells and may be responsible for this maximum life span: they showed that structures at the ends of chromosomes called telomeres become shorter each time a cell divides, and that an enzyme called telomerase can restore some of that length up to a point (Grieder and Blackburn 1985).

Evolution of Aging

Aging has not always been with us. The first organisms, which consisted of only a single cell, likely did not age. Scientists assume that they reproduced by simple cell division that resulted in two equal offspring (Ackermann et al. 2007). Over time, those organisms must have accumulated damage to their molecules and organelles, and the two daughter cells from cell division must have received approximately half of the damaged material. The two equivalent daughter cells were able to deal with the amount of damaged material that they inherited. If not, the line would have died out. This is the reasoning behind the notion that unicellular organisms do not age.

The alternative to the early equivalent division is an asymmetric situation in which the damaged material is sequestered into one daughter cell and the other daughter cell receives the newly produced proteins and organelles. That outcome would require a more complicated mechanism to ensure the separation of old and new components. At some point in evolution, that additional mechanism did appear. Indeed, more recent experiments show that the seemingly random sorting of components into the two daughter cells might not be truly random (Watve et al. 2006; Clegg, Dyson, and Kreft 2014; Yang et al. 2015). With an asymmetric distribution of the damaged material, one cell would get most or all of the damaged material. That cell might be called the "mother" cell. The other cell would get all or most of the new material. That cell might be called the "daughter" cell. The mother cell would likely accumulate additional damaged material after more division and would eventually die once the cell's systems were overwhelmed by too much damage. The daughter cell would get a new lease on life at each division.

For some time, scientists assumed that this evolutionary step took place with the eukaryotes. Bacteria, which are prokaryotes, were thought to reproduce by simple division to produce equivalent daughter cells. However, more recently, bacteria have been found that have asymmetric division: producing a mother cell and a daughter cell. Over multiple generations, the mother cell reproduces less often and eventually dies. These observations support the notion that asymmetric division and aging (in the sense of a decrease in reproductive ability and an increase in intrinsic mortality) began in simple organisms such as bacteria and yeast and is carried on in eukaryotes, including humans. The ability to segregate damaged components into one daughter cell would likely be favored by natural selection.

Lower organisms display several types of aging. The protozoan *Tokophrya* reproduces by an internal budding process that leaves it without a mechanism for disposing of wastes. This mechanical senescence ensures that the organism will age and die. *Paramecium* and *Tetrahymena* experience replicative senescence—after reproducing asexually for some time, they lose that ability—but they can get a new lease on life by undergoing a round of sexual reproduction.

Sexual reproduction brought new facets to aging. The exchange of genetic material that happens in sexual reproduction is certainly an advantage—it increases diversity, which in turn increases the likelihood that organisms will be able to adapt to change.

It's attractive to think that old people age and die to spare scarce resources for the benefit of a younger generation. Many cultures revere their older members for their accumulated experience and wisdom. However, at some point, the resources needed to maintain infirm and unproductive older adults outweigh that benefit. Thus, the final service that the old people give to their community is to get out of the way of the next generation. Death would be a noble deed and final gift to our posterity. We would be sacrificing ourselves for our children and grandchildren, and that might give our life and death meaning.

Yet as warm and comforting as that explanation might be, it is simply wrong. Leslie (2001) sums it up well: "So why do our bodies gradually fail? We might be tempted to think of aging as a bit like mandatory retirement. From this perspective, aging serves to cull the old and make way for the young, thus promoting the survival of the species. But this naive argument, which still appears embarrassingly often in the literature, fails at every level."

The first and most critical level involves evolution. The force behind evolution is natural selection, and the acid test of any theory of aging is that it must account for natural selection (Kirkwood and Austad 2000). Genes that benefit fitness early on in an individual will be favored because those individuals will reproduce

more successfully. Genes that hinder fitness early on will be selected against because those individuals will reproduce less successfully. Genes can be selected for
only if they affect the organism before reproduction. Aging and death generally
occur well after the reproductive years, and the force of natural selection is lost
later in life. Thus, the "altruistic" theory of older people sacrificing their lives to
benefit younger generations cannot be correct.

Never fear. There is no shortage of theories to explain aging. Medvedev
(1990) created a list of 300 of them. And while none has proved to be the complete answer, three have emerged as good explanations of how aging might have
evolved: mutation accumulation theory, antagonistic pleiotrophy, and disposable soma (Robins and Conneely 2014; Ljubuncic and Reznick 2009). These theories are not mutually exclusive, and of course other factors may well be involved.

Mutation Accumulation Theory

The mutation accumulation theory of aging was proposed in 1952 by Sir Peter
Medawar (who would later win the Nobel Prize in Medicine in 1960 for his work
on immunology that provided the basis for organ transplantation). The theory
states that the deleterious effects of genes that affect an organism early in life have
a large negative selection value because they hinder reproduction. However,
genes that are deleterious later in life have less or even no negative selection
value—the organism has already reproduced, so genes with later effects are essentially meaningless for evolution.

Medawar (1952) suggested that aging was caused by mutations that were detrimental only later in life. A good example is Huntington's disease, where a mutation in the gene for the protein huntingtin (the function of which is unknown)
results in a protein with a long string of the amino acid glutamine that is not present in normal huntingtin molecules; once the repetitions reach a certain number
(about thirty), disease results. It is an autosomal dominant disease—having one
allele for the disease means you will eventually get the disease. Huntington's disease usually manifests after about age fifty, well after the reproductive years.
A person with an affected parent has a 50 percent chance of inheriting the mutant gene and getting the disease. Unfortunately, for many years there was no test
for the defective gene, and by the time the disease manifested, the patient has
long since had children and might have passed on the toxic gene. Today a DNA
test is available to determine if a person is a carrier of the defective gene before
they consider becoming parents.

Medawar assumed that a number of mutations that operate similarly to the
one that causes Huntington's disease are spread throughout the human population. They manifest in later life, but instead of causing a specific disease, they

cause a series of effects that we have come to look upon as "normal aging." Furthermore, those mutations likely occur in genes involved in maintenance of the fully grown organism, rather than genes involved in growth and development. Cell division, for example, would only occur to replace damaged cells or those that normally turn over rapidly (e.g., blood, skin, or intestinal cells). In this theory, aging could be avoided by removing or neutralizing these mutations or by treatments that remedy the effects of the mutations.

The mutation accumulation theory is not without its problems. For technical reasons, it works only if the mutations have essentially no effects on fitness. Most animals probably do not live to old age in the wild, but it is hard to believe that mutations that affect speed or agility in the young would not also affect older animals. Medawar's theory correlates well with the different ages of sexual maturity of mammals. However, he did not include non-mammals. And the theory does not account for animals that die immediately after reproduction. For example, salmon return to the same river from which they spawned years before to reproduce, and immediately after spawning, they die. This is referred to as semelparity, and it occurs in a number of other species, including some mollusks, butterflies, cicadas, many arachnids, octopus and squid, and mayflies. An accumulation of mutations cannot explain this seemingly genetically programmed death. Finally, many genetic diseases have causes more complex than a mutation in a single gene; gene programs involve multiple genes that turn on and off at different times throughout growth and development.

Medawar's mutation accumulation theory, at least in its initial formulation, seems to be too simple to account for aging. However, it is difficult to make a final conclusion about the basic idea. Charlesworth (2001) suggested a modified mutation accumulation theory that is a better fit with the mortality data, and recent computer analyses support this theory (Reynolds et al. 2007).

While the mutation accumulation theory is widely respected, there are some who are critical of it. Some experimental studies have focused on positive pleiotropy (pleiotropy refers to the ability of a single gene to have more than a single effect, and pleiotropic effects are common in biology), which may actually improve fitness across ages and even slow aging and increase life span (Maklalov, Rowe, and Friburg 2015). These results suggest that the evolution of aging is even more complicated than thought, and more research will be needed to come to a conclusion.

Antagonistic Pleiotropy Theory

The antagonistic pleiotropy theory of aging was suggested by George Williams (1957). As noted in the discussion of pleiotropy, a single gene might affect multiple traits. According to the theory, evolutionary pressures favor gene mutations that have benefits earlier in life, and thus promote reproductive success, over mutations that promote long life. And mutations that produce benefits earlier in life but cause problems later in life cannot be selected against, because their effects don't show up until after the reproductive years are over. So in Williams's theory, a gene that has multiple effects favoring fertility early on is selected for even if those effects cause problems later on. Those late-manifesting problems are what we call aging.

In the wild, this might have no practical implication, since most organisms die from other causes before late-acting detrimental genes are activated. However, there are examples in nature suggesting how Williams's theory might work. For example, the long tailfeathers of the peacock help the males attract females and so encourage successful reproduction early on. However, later in life, those same feathers might hinder the bird's escape from a predator (Zahavi 1975). Another example might be that having a limit on the number of times a cell can divide suppresses tumors early on, but in older organisms the resulting senescent cells might allow cancers to develop more easily. Yet another example might be sex hormones that increase sex drive and reproductive efforts early on but might cause hormone-linked cancers (prostate and ovarian) later in life. Martin (2007) suggested there might be paradoxical antagonistic pleiotrophy, in which genes might be beneficial in old age but detrimental earlier; however, there would be selection pressure against genes like that. An example would be plasminogen activator inhibitor-1; higher levels of this substance are found in people over 100 years of age, but when these higher levels appear in younger people they are linked to a higher incidence of heart attacks. And Rose (1984) showed that in fruit flies (*Drosophila*), individuals that reproduce later also age more slowly and vice versa.

Disposable Soma Theory

Thomas Kirkwood and Robin Holliday (1979) suggested that when resources are limited, organisms must make compromises about how they allocate those limited resources to various types of bodily functions, including growth, reproduction, and repair. If they use too many of their limited resources on repair, for example, reproduction doesn't get enough, and their offspring will suffer and perhaps not reach reproductive age. Thus, reproduction is favored and receives most of the resources. With fewer resources available for repair, however, cellular damage may not be fixed. Cells are damaged, telomeres become shorter, and mutations accumulate. Eventually, senescence and then death set in. A variation on this theme asserts that time can replace limited resources. In other words, organisms that live longer have more time to devote to high-quality repairs than short-lived organism. Drawing on systems analysis, Kriete (2013) expanded this concept to include traits such as "robustness."

There is considerable evidence for this theory. First, there seems to be an inverse correlation between size and life span in many animals, suggesting that resource trade-offs are made between growth and maintenance. Small dogs typically live fifteen to twenty years, but the largest dogs live only six to eight years. In humans, individuals with Laron syndrome, a form of dwarfism, live longer than average, and centenarians tend to be shorter of stature than those who die younger. And animals that reproduce earlier or more tend to live shorter lives, again suggesting trade-offs between reproduction and life span.

There are also arguments against the theory. The biggest criticism is that the theory suggests no mechanisms. Another criticism is that calorie restriction is well known to extend life span in a number of organisms, and at first it might seem that this fact is at odds with the disposable soma theory. After all, calories are a resource, so if we limit the supply of them, we wouldn't expect life span to increase. However, reproduction is also reduced when calories are severely restricted, so the life span benefits of caloric restriction are not a complete argument against the theory. And naked mole rats do not fit the theory: they do not seem to age, as we've seen, yet those that reproduce live longer than those that do not.

A more serious criticism is that women invest much more in reproduction than men, and yet they tend to live longer than men (Blagosklonny 2010). Furthermore, longer-lived grandmothers provide support for younger women and their children, helping to ensure that the next generations survive (Hawkes 2003). The counterarguments are that menopause makes women's reproductive time shorter than that of men. Also, men invest more in growth, which might help account for their shorter life span. Thus, the theory remains a possible explanation for the evolution of aging.

Modern Theories of Aging

The three classic theories of aging provide a powerful basis for thinking about the evolution of aging. Recently, several additional theories or variations on the classic theories have been suggested. Like the classic theories, they are not mutually exclusive. They are also not completely compelling.

Reliability Theory

Gavrilov and Gavrilova (2003) point out that many cells and organs seem not to age at all. For example, neurons do not seem to have increased risk of death as they get older. They question how a system that includes non-aging parts can age. Their mathematical models also show that some animals seem to stop aging after a certain point, including humans over 100 years old. Their reliability theory is a variation on a theory that has been used successfully in engineering and manufacturing. In biology, it suggests that humans are complex systems with many parts and multiple redundant systems. These ensure robust health early on, but as we grow older, those systems begin to fail one by one, and that results in aging. So aging is a trade-off. For example, redundancies in systems that involve proteins may enable the organism to maintain good health early on, but later those systems fail, as in the cases of neurodegenerative diseases associated with the accumulation of damaged proteins, such as Huntington's disease, Alzheimer's disease, and Parkinson's disease.

Inflamm-Aging

Another theory of aging, called inflamm-aging, posits that chronic, low-grade inflammation is a cause of aging (Franceschi et al. 2000). The immune system is complex and its various parts are highly regulated to keep it in balance. Our adaptive immune system, which relies on cells that can learn to identify and destroy pathogens, becomes less effective over time, as the number of new or "naive" B and T lymphocytes that can recognize and fight infection decreases with age. On the other hand, the innate immune system uses specialized chemical substances called cytokines (e.g., interleukin-6, tumor necrosis factor alpha, and C-reactive protein), among other things, to defend the body by causing inflammation.

In this theory, the innate immune system is activated by the types of chronic infections that plague the aged, the increased amounts of cell debris that accumulate over the years, and reactions to the gut microbiome, among other things (Basisty et al. 2020). Against this background of low-level inflammation, second hits can push the system out of balance and cause disease conditions, including Alzheimer's disease, atherosclerosis, type 2 diabetes, and chronic heart disease. In fact, research has shown that many age-related diseases have a long subclinical phase, and over time inflammation takes a toll. These findings are in keeping with the antagonistic pleiotropy theory of aging.

DNA Damage Theory

This theory suggests that aging results from an accumulation of unrepaired damaged genes. Each human cell is estimated to experience 70,000 incidents of damage per day (Lindahl and Barnes 2000), including mistakes during replication or damage from chemicals or UV light. These incidents are of many types, including deletions, insertions, substitutions, cross-linking with other nucleotides or proteins, chemical modifications, single- or double-stranded DNA breaks, and chromosomal rearrangements (Moskalev et al. 2012). Seventy thousand might seem like a huge number, but it is important to remember that each cell contains 3 billion base pairs of DNA. (Interestingly, the rate of these changes is fairly constant and can be used to estimate evolutionary times.) Organisms have evolved repair mechanisms to correct at least part of the damage; nevertheless, changes accumulate over time. It is tempting to connect that damage with senescence (Freitas and de Magalhães 2011).

The central nervous system may be particularly susceptible to DNA damage. Brain tissue is postmitotic (Maynard et al. 2015), meaning that cells in the brain no longer divide to create new cells that might replace diseased or damaged cells. Also, damage to the DNA of our mitochondria has been linked to neurodegenerative diseases.

There is evidence for this theory. Mutations that disrupt the genes for DNA repair accelerate aging in mice. In humans, progeria, as mentioned in Chapter 1, is a disease that causes accelerated aging, and it is associated with defects in DNA repair. Furthermore, certain mutations, called single-nucleotide polymorphisms, that upregulate the expression of DNA repair genes are associated with lengthened life spans. And centenarians also have higher levels of DNA repair enzymes than the average of those of ages sixty-nine to seventy-five.

Programmed Aging

Several aspects of biological systems might be programmed to age. Specific sets of genes might turn on or off, and the affected systems could include hormones that control biological clocks. The immune system might lose its effectiveness over time to allow diseases.

Wear and Tear

This is a very old theory. It was first suggested by August Weismann in 1882, and it has a lot of appeal to many people because it is easy to understand the analogy of the human body with machines that simply wear out. This theory suggests that aging might be due to simple wear and tear on an organism's systems that eventually overwhelms the body's repair mechanisms. The modern update on this theory suggests that proteins, lipids, and nucleic acids might be damaged from cross-linking or oxidative damage due to free radicals that are normally produced during metabolism. And DNA might be damaged in the ways we've just discussed.

Phase Separation Problems

Alberti and Hyman (2016) suggested that cellular aging might be due to dysfunctional phase separations. The cytoplasm—that is, the contents of a cell, excluding the nucleus—is organized into membrane-bound organelles and membraneless compartments that involve phase separation. Phase separation is sensitive to changes in concentration, pH, and energy levels. When these separations go awry, cell death can result.

Molecular Genetics

Advances in modern genetics have shown that mutations in some genes significantly extend the life span of some organisms (covered in detail in Chapter 7). Single gene mutations extend life span in some organisms. An example is the *daf-1* gene in nematodes and other genes in fruit flies. Genomic studies and studies of total gene expression can show if there are differences in individuals at different ages. The studies that have been done so far suggest that mice and humans age differently (Lund et al. 2002; Zahn et al. 2007). It is not quite clear what all of these findings mean for humans, but they are intriguing.

Pal and Tyler (2016) suggest that epigenetics—changes in gene function that do not involve changes in DNA sequence—may have a significant role as a mechanism in aging processes. The good news is that epigenetic modifications are controlled by enzymes, and thus they are reversible. That reversibility points to the possibility of controlling aging by controlling epigenetics (López-León and Goya 2017). These results support the concept that epigenetics is deeply involved in aging and might be a key to the reversal of aging.

Can We Do Anything About Aging?

First, we should remember that a lot has already been done about aging. The average life span for humans has increased significantly in the last century. In addition, health span—the time we spend in good health—has been improved for many people. So people are living longer and also healthier lives. Improvements in nutrition, access to healthcare, and medical advances are largely responsible for this. Although some factors have slowed the increase in life span in the

United States recently, there is no reason to believe that new discoveries will not continue to lengthen our lives.

Second, researchers have come up with plenty of valuable theories and lots of experimental data that can be used to investigate ways of extending our lives and decreasing the impacts of aging. These in turn generate new theories and suggest new ways to test these theories and evaluate data.

One thing is clear: genes are important in aging and in how long we live. However, how genes actually work in aging isn't so obvious. As noted earlier, natural selection favors genes that promote reproduction, but there are no mechanisms to select for genes that favor long life (Kirkwood 2008). The involvement of genes in life span is likely to be very complicated, making it harder for us to figure out how we might intercede to encourage genes promoting longer life.

It also seems accurate to say that aging appears to be caused by our limited ability to maintain our cells (Kirkwood 2005). As a result, damaged proteins and organelles tend to accumulate over time and cause disruption of critical cellular activities.

Nelson and Masel (2017) developed a model of aging and used it to test various outcomes. Their findings are consistent with theories of cell senescence. Senescence is selected against in unicellular organisms by competition among cells. That is not the case in multicellular organisms; because their existence depends on cooperation among cells, competition between the cells of a single organism is mitigated to ensure the survival of the whole organism. That cooperation leads to the accumulation of senescent cells.

Others question Nelson and Masel's conclusions. They believe that aging is not inevitable. For example, Mitteldorf and Fahy (2018) point out that the aging described by Nelson and Masel is not human aging. In fact, there are many forms of aging in nature, and few of them conform to Nelson and Mazel's paradigm. Mitteldorf and Fahy further note that most of the major causes of aging are amenable to treatment and that research on aging is likely to yield results. Not incidentally, they also noted that Nelson and Masel's conclusion that aging is inevitable got a lot of play in the press, and perhaps that might discourage worthwhile research.

Aging as a Disease

In an interesting paper entitled "How Can Aging Be Thought of as Anything Other than a Disease?," Caplan (2015) summarizes the discussion about aging as a disease in this way: "Aging exists solely as a consequence of evolution's lack of 'foresight.' It is simply a by-product of selective forces that work to maximize the organism's immediate chances of reproductive success. Senescence has no function. It is simply the inadvertent subversion of organic function, later in life that results from favoring reproductive advantage early in life."

Is aging a risk factor for disease, or is it a disease itself? Certainly accelerated aging (e.g., progeria) is considered a disease. For many, though, aging is seen as a "natural" process, whereas diseases are "unnatural." On the other hand, a number of the manifestations of aging are considered diseases, such as osteoporosis, atherosclerosis, cancer, arthritis, cataracts, type 2 diabetes, hypertension, and many neurodegenerative diseases. A significant advantage of labeling aging as a disease is that it would be a rejection of the fatalistic view that aging is normal (Hull, Bjork, and Roy 2015). Even if aging is not viewed as a disease, it might still be an appropriate target for medical intervention or lifestyle interventions such as exercise and nutrition, as is the case with other non-disease conditions such as pregnancy.

That said, *should* aging be viewed as a disease? That's a more complicated question than it might at first seem. First, the term *disease* is difficult to define. The World Health Organization (WHO 1946) defined health as a "state of complete physical, mental, and social well-being and not merely the absence of disease or infirmity." How does aging fit in with that?

Next, what is considered "normal aging" and what is considered "disease" changes over time. For example, Alzheimer's disease, hypertension, and osteoarthritis were once considered to be normal aspects of aging; now we recognize them as disease states that can and should be treated. We also recognize that aging is the major risk factor for these diseases. That is part of the reason that aging itself is being considered by some as a disease that might be amenable to treatment (Bulterijs et al. 2015).

What about the physiological decline that accompanies aging? Is that a disease? At this point, most scientists and physicians would likely say no. However, a closer look reveals that the issue may not be so clear-cut. Just as we now consider Alzheimer's a disease state, many of the other conditions associated with aging might be thought of as disease states if considered in isolation. In addition, an increasing number of the symptoms of aging have a specific cause and are understood down to the molecular and cellular levels (López-Otín et al. 2013).

The traditional view is that aging is a normal and natural process—the body just wears out. As a part of this process, older people are naturally susceptible to diseases of old age, such as Alzheimer's disease and various cancers. Because this is a normal and natural process, that view continues, we should simply accept that outcome Some have even made the argument that it is the duty of the aged to die and make available resources for the next generation.

However, this position is difficult to maintain. We invest in a vast research effort to understand, prevent, and cure the diseases of the aged. Neurodegenerative diseases, heart disease, and cancer cause great amounts of suffering and incur enormous costs at all levels of society. No one would advocate for ending work to relieve that suffering.

Of course, attacking these diseases is not the same as attacking aging itself. A more ambitious strategy for ending aging is to look for interventions that control aging before it leads to disease. This strategy has been successful in some model organisms. For example, the life span of a worm, *Caenorhabditis elegans*, can be greatly increased with a substance called resveratrol, and so it might be possible to find such interventions for humans. Of course, this approach would only work if there actually is a central control system for aging.

The main theories of aging give some hints about how to proceed. For example, therapies might increase the cells' repair mechanisms. We might look at some of the molecular mechanisms that are shared between the diseases of aging and aging itself, which would suggest that those diseases are an accelerated form of aging (Franceschi et al. 2018). As evidence, researchers point out that at one extreme of the aging spectrum are people with progeria and Werner's syndrome, who age very rapidly. At the other extreme are centenarians, who rarely have any aging-related diseases and display lower levels of the various markers of inflammation. Perhaps that means we should look more closely at chemical signaling networks, such as the IGF-1 and mTOR systems, that are involved in both diseases and aging (Niccoli and Partridge 2012), or at ways to block inflammation.

While many studies have looked at centenarians, they have yielded relatively little information about what helps this group live so long. Erikson et al. (2016) took a different tack, comparing a cohort of individuals who were between 80 and 105 years of age and had no recognized diseases with a group of typical Americans in their seventies and eighties. While they found no single factor that guaranteed a healthy life, they did find some clues; perhaps that suggests we should expand our comparisons to young adults as well. Belsky et al. (2015) did just that. They examined 954 young people in their twenties and thirties for several system functions, including pulmonary, periodontal, cardiovascular, renal, hepatic, and immune, and found that some were already aging more rapidly than others. These results show that aging research in younger subjects could be beneficial.

Another intriguing experiment involves cell reprogramming. Reprogramming of cells has mostly been thought of as a potential way to repair or replace damaged tissues or organs. However, Ocampo et al. (2016) tried a different strategy. They briefly treated mouse cells used as a model of premature aging (Hutchinson-Gilford progeria) with the Yamanaka factors (Oct4, Sox2, Klf4, and c-Myc), transcription factors that can reprogram differentiated somatic cells into stem cells. The treated mouse cells did not change from adult cells to stem cells, but the treatment did improve several markers of aging, including DNA damage, senescence, and production of reactive oxygen species. In addition, it reset the levels of modifications to histones normally associated with aging.

What all this shows is that there are many possible approaches to aging, and while there are many practical challenges, there are also a number of promising directions for research.

5

Dying

Death must be so beautiful. To lie in the soft brown earth, with the
grasses waving above one's head, and listen to silence. To have no
yesterday, and no to-morrow. To forget time, to forgive life, to be
at peace.

—Oscar Wilde

Death is not always as gentle and poetic as Oscar Wilde described. People die
from many causes. Some die peacefully in their sleep. Others die from a gunshot wound to the head, a heart attack, a traffic accident, or an explosion. The
transition from life to death may take only moments; a completely healthy and
vigorous person may not even realize that anything has happened before they
lose consciousness. Other deaths are more protracted, such as from progressive
neurodegenerative diseases or cancer. And there is everything else in between
these extremes.

What happens near and at the moment of death? Here this question does not
concern sudden death, for that needs little explanation. Rather, we want to look
at what happens when death occurs slowly. What happens in the hours and days
before death? For patients who are suffering a slow death, are there signs that
death is near? The clear answer is yes, and we will list some here.

Clinical Signs of Approaching Death

Knowledge of the clinical signs that signal the beginning of the dying process is
valuable for physicians. Accurately predicting death in terminal patients is difficult but very important. These signs indicate the beginning of irreversible steps
toward death, and that can help medical personnel and caregivers to determine
appropriate courses of action, such as transfer to hospice and when to discontinue aggressive treatment.

It is important to distinguish changes that typically occur as a part of aging
from those that are a part of dying. What we might assume is a normal part of
aging might in fact be a treatable conditions. Diehr et al. (2002) looked at ten

The Biology of Death. Gary C. Howard, Oxford University Press. © Oxford University Press 2021.
DOI: 10.1093/oso/9780190687724.003.0005

variables associated with aging and dying: self-rated state of health, depression, activities of daily living, instrumental activities of daily living, scores on the Mini–Mental State Examination, body mass index, days spent in bed, hospitalization, blocks walked per week, and walking speed. Most of these measures declined with age, but they seemed to decline more rapidly as death approached. The researchers concluded that sudden declines in any of these should be taken seriously.

Gerontologists use the term *terminal decline* to describe the gradual loss of mental or physical capabilities as death approaches and *terminal drop* to describe a more rapid loss. Both of these terms could apply either to aging generally or to the process of dying, and it isn't clear if they are distinct events or simply points on a continuum. However, in general, *terminal decline* describes the gradual decline in capabilities that accompanies aging, while *terminal drop* is a much more rapid loss of these capacities and tends to be associated with the approach of death. Not everyone experiences such a decline. MacDonald, Hultsch, and Dixon (2011) looked longitudinally at the decline in mental function in aging patients. They found a gradual decline in mental abilities that accelerated somewhat nearer to death. However, they did not see the precipitous decrease that would indicate a terminal drop.

Hui et al. (2014) carefully examined ten signs that seemed to signal that the end of life is near for cancer patients:

- The patient stops breathing for periods (apnea)
- Breathing is abnormal and alternates between periods of increasingly deep breathing and periods of shallower breathing (Cheyne-Stokes respiration)
- Distinctive sounds occurring when the patient can no longer swallow or cough up saliva to clear the throat (death rattle)
- Problems with swallowing liquids (dysphagia)
- Decreased consciousness
- Extremities turn bluish (peripheral cyanosis)
- Radial artery lacks a pulse
- Breathing in which the jaw drops (respiration with mandibular movement)
- Urine output is less than 100 ml over a period of twelve hours
- Palliative performance scale result of less than 20 percent (a measure of a patient's ability to complete daily activities)

Scales are often used to determine the status of cancer patients. The Karnofsky scale (Péus, Newcomb, and Hofer 2013) uses a score of 0 to 100 points. The WHO scale (Oken et al. 1982) is much simpler. Children are often scored on the Lansky scale (Lansky et al. 1987), which depends more on adult observations.

Hui et al. (2014) placed the signs into two groups. The early signs included decreased performance status, change in oral intake, and change in level of consciousness. These early signs were observed relatively frequently, but they were not very specific; they could not allow the pinpointing of death to within three days. However, they are still useful as indicators that the patient is deteriorating and that death may occur within weeks. Later signs appeared in only some patients—for example, death rattle was observed in 66 percent of patients—but when they did occur, they were observed specifically only in the last days. These signs were much more effective in predicting death within three days. However, their absence was no guarantee that death would not occur soon.

One important caveat is that people are different, and so any given sign may or may not be seen in a specific case. Still, there are some generalities we can observe.

First, appetite decreases. The body needs less energy, and so the patient slows or stops eating and drinking. They may sleep more as their metabolism becomes slower, and they have less energy. They might become less social, and they might not be able to sustain a conversation. Their vital signs will change: blood pressure drops, the rate of breathing slows, breathing may start and stop with long pauses (called Cheyne-Stokes respiration), the heartbeat becomes weaker and irregular, and the urine might become brown or rust-colored as the kidneys begin to fail. Because they are eating and drinking less, their needs for elimination may be less or none at all. Their muscles will become weaker. Their temperature may drop, a sign the body is rationing blood circulation so that the internal organs get the lion's share. Their hands and skin may be cool to the touch, and the skin may become pale or bluish. Their mental state will change: they may seem confused or they may experience hallucinations. The senses decline, with hearing thought to be the last sense to be lost. Sadly, they may experience pain.

When death occurs, some of the signs are obvious: There is no pulse. They stop breathing. The limbs are limp. The eyes become fixed. The bladder and bowel may release their contents.

In one retrospective study (Witkamp et al. 2015), researchers asked 249 relatives to complete a questionnaire on the final twenty-four hours of the patient's life. On average, patients suffered from several symptoms: dry mouth, loss of consciousness, fatigue, and difficult breathing. They also had psychological symptoms, such as feeling powerless, sad, anxious, and worried. Slightly more than one-third of patients were at peace. About half had indicated their preferences for treatment at the end of life. About one-third had noted where they preferred to die, and of those, most wanted to die at home.

What Happens at the Moment of Death and
Shortly Thereafter

While death stops the processes of life, they do not all stop at the same time. A series of events begins. The pulse disappears. Breathing stops. In the first hour after death, all of the muscles in the body go limp. The eyelids lose tension, the pupils dilate, and the joints and limbs are loose and flexible. A few minutes after the heart stops pumping, the blood begins to drain from the smaller blood vessels, and the skin becomes pale in the upper parts as the blood begins to settle and pool at the lowest points. The body begins to cool; the rate of cooling is about 2°C in the first hour and 1°C for each hour after that. The sphincters relax, and urine and feces are released. Between two and six hours after death, the blood continues to settle to the lowest points in the body, which become reddish purple. The skin may look like it is bruised. In the third hour, the body begins to stiffen as the muscles go into rigor mortis. This biochemical reaction involves calcium ions and the contraction of the myosin and actin fibers that make up the muscle tissues. Rigor begins with the eyelids, jaw, and neck and spreads over the next hours to the rest of the body, ending with the fingers and toes. Rigor is greatest about twelve hours after death. After twelve hours, rigor begins to lessen due to further biochemical reactions within the muscles. By about forty-eight hours after death, rigor has disappeared. Infants do not demonstrate rigor.

Interestingly, although the heart and lungs stop their activity at death, not all other organs die at the same time. For example, there is evidence that brain activity spikes just before death. It isn't clear what this means, if anything, in terms of consciousness. Borjigina et al. (2013) noted a surge of synchronous gamma waves within thirty seconds after a cardiac arrest in rats. The activity in the dying brain was greater than that in normal healthy brains. Neuronal death is thought to involve massive increases in the extracellular concentration of neurotransmitters (especially glutamine) and loss of the supply of oxygen, glucose, or both (Martin et al. 1994). The membrane potential changes, levels of glutamate become neurotoxic, and the ion gradients change dramatically.

The changes that occur in the brain during the dying process can be examined by looking at the cortical electroencephalographic (EEG) patterns (Schramma et al. 2020). In rats suffering cardiac arrest, the EEG amplitude declines rapidly and is followed by a surge of high-frequency gamma oscillations. In another rat model, a large slow polyphasic wave suddenly appeared about a minute after decapitation. This wave, called the "wave of death," is thought to represent the death of neurons. In other models, the wave of death could be reversed if the oxygen supply was restored within a few minutes. These changes are not well understood, but if they could be reduced to biomarkers, they could be very useful for helping to determine the best timeline for treatments.

Do All of Our Cells Die at the Same Time?

When a person dies, not all of the cells die at the same time. Some continue to live for hours or days, and continue to carry on their normal functions. Genes are turned on or off, and transcription of RNA and translation of proteins continue. Counterintuitively, some genes increase their expression.

One team of scientists examined over 7,000 tissue samples from 540 donors (Ferreira et al. 2018) and found that each tissue type dies at a different rate and with different genes turned on. Another study showed that about 99 percent of all genes stop transcribing RNA at time of death, but 1–2 percent increased transcription. In zebrafish and mice, some genes increased activity immediately after death, but others showed increased activity twenty-four, forty-eight, or ninety-six hours after death (Pozhitkov et al. 2017). The authors of the study called this time the "twilight of death"—the person or animal isn't living, but also isn't completely dead either. The genes that were turned on were a mixed bag. Some were not surprising; increasing the expression of genes for wound healing and oxygen supplies makes sense. But others are harder to understand. For example, some genes that were turned on were those involved in embryonic development and cancer. However, although the types of genes that were turned on did not seem to fit any pattern, the order in which they were turned on did show a pattern: the same genes were turned on at the same time. In fact, these sequences of activity are so regular that they can be used to determine the time of death (Hunter, Pozhitkov, and Noble 2017).

Of all the organs in the body, the brain is the most vulnerable to inadequate oxygen supply (hypoxia) and inadequate blood flow (ischemia). Key cells (neocortical pyramids of layers III, IV, and V, hippocampal CA1 pyramidal cells, striatal cells, and cerebellar Purkinje cells) are irreversibly injured within ten minutes of loss of blood circulation. The consensus is that when death occurs, blood flow ceases to the brain, and oxygen supplies are rapidly used up. Electrical activity in the brain spikes for a few tens of seconds. After that, the activity decreases, and then ceases. The stoppage spreads slowly and can take tens of minutes to cover the entire brain. These phenomena have not been examined carefully, but a better understanding of the processes would help clinicians predict how much time they have to try to restore blood flow before irreversible damage occurs.

Pani et al. (2018) followed the loss of brain activity in a male macaque monkey that was sedated and given a bolus of pentothal sodium to cause cardiac arrest. About twenty minutes after the cardiac arrest, brain activity reemerged and remained for 120 minutes. This experiment involved only one animal, but it did carefully document brain activity for two hours after the end of heart activity.

Dreier et al. (2018) also studied brain activity after death. They followed events in terminal patients after life-sustaining treatments had been withdrawn.

All of the patients had do-not-resuscitate orders. They monitored brain activity as the patients died and found that even five minutes after death, the brain used its small reserves of energy to remain active. As the brain runs out of reserves, the cells shut down. But even then, the possibility remains that they might still be reactivated by a fresh blood supply. How long they can be shut down and still be able to be restarted is unclear.

A recent experiment complicated our vision of death. It showed that some functions of pig brains that had been dead for up to four hours could be reestablished (Vrselja et al. 2019). They perfused the brain with an acellular, hemoglobin-based fluid. The brains showed no electrical activity, but they did display several features of living brains. Those features included appropriate cytoarchitecture, slowing of cell death, spontaneous synaptic activity. As this line of experimentation continues, it is likely to have significant consequences for organ donation and the removal of life support from patients.

Near-Death Experiences

In dying, there is an instant in time when a threshold is crossed, and in that instant, the person changes. One moment, he is an individual with a personality, memories, and characteristics that define him and his life. The next, he is gone forever. The spark of life, whatever it is, is gone.

Little is known about exactly what happens during dying. However, from time to time, stories appear in the popular press about those who had a near-death experience (NDE). There are many accounts of these experiences. A quick search yields scores of accounts, both written and on video. The experiences occur in response to many different clinical events, including cardiac arrest, septic shock, electrocution, traumatic brain injuries, strokes, near-drowning, and more. Some patients have them during terminal illnesses, others in conjunction with severe depression. The patients are declared "dead" by the usual criteria: they have no heartbeat or respiration, and their pupils are fixed and dilated. However, later they are resuscitated by medical intervention and survive this experience. The common interpretation is that they were dead but somehow managed to return to life by their own will or because it was not "their time."

Were these people really dead? Are their experiences representative of the normal experience of dying? Does what they report give us any hints about the afterlife? That isn't so clear. Only a small percentage of patients are "brought back" from dead, and of those, an even smaller number report a NDE. Controlled experiments are rare and very difficult, and so conclusions are limited. Nevertheless, we might glean a few insights from some of these reports.

Individuals who survive these experiences report a variety of events. Not everyone who has a NDE can describe the experience. Some are simply too ill to speak. Others do not remember anything. Those who can recall the events have different stories, and those are influenced by culture. Interestingly, and perhaps tellingly, quite a few report similar experiences (Martial et al. 2017). For example, they believed that they were dead. They had a sense of well-being. They often could see their own body as if they were outside of it observing events as they were occurring. They seemed to be moving toward a bright light. Some entered the light. Finally, many saw their life "flash before their eyes." They were able to speak to other beings, and some of those beings were long-dead relatives. Many of those who have had a NDE report that time seems to slow down or become distorted. Wittman et al. (2017) asked 196 survivors if time had seemed to either speed up or slow down, and 127 (65 percent) answered yes. Their experiences were different, but they sensed that they lost track of time in those minutes.

Do these experiences and especially these common features have any meaning? Are they a glimpse of life after death, hallucinations generated by a dying brain, or the expected results of physiological processes? Many possibilities have been suggested. Some are spiritual, some are psychological, and others are physiological.

Van Lommel et al. (2001) came to a different conclusion. They completed a prospective study (a study in which patients are enrolled before they have the disease or incident) of cardiac patients and wondered why more patients did not report a NDE. If purely medical factors (such as hypoxia) accounted for NDE, then many more patients should have reported the experience. Significantly fewer survivors had such an experience than have been found in retrospective studies. Furthermore, the symptoms were not explained by many variables, including the duration of cardiac arrest or unconsciousness, types of medications used, or the level of fear of death that the patient had before cardiac arrest. They also found that the frequency of reports of NDEs was influenced by how such experiences were defined. Other factors included the age at which the patient first survived a heart attack, the number of cardiopulmonary resuscitation events in the hospital, whether the patient had experienced a previous near-death event, and the presence of memory problems. The authors suggest that additional research be carried out on specific aspects of the NDE.

There is good evidence that physiological changes that occur upon loss of circulation may cause the effects reported in NDEs. For example, a prospective study (Klemenc-Ketis, Kersnik, and Grmec 2010) in the three hospitals in Slovenia examined fifty-two patients who survived cardiac arrest. They looked at the effects of changes in oxygen levels, carbon dioxide levels, and sodium and potassium ions in patients who had their NDE outside of a hospital. In addition to the blood chemistry assays, they administered Greyson's NDE scale (Greyson

1983). They found that high levels of carbon dioxide and potassium might be involved in causing the reported elements of NDEs. In these cases, the anoxia (an extreme form of hypoxia, or inadequate oxygen supply) produced symptoms that were very similar to those reported by individuals who had experienced a NDE. The researchers confirmed their hypothesis. Patients with the highest levels of carbon dioxide had more NDEs than those with lower levels of carbon dioxide.

Other effects could also be explained physiologically. During a heart attack, endorphins are released that can result in hallucinations. Other drugs produce hallucinations. For example, ketamine produces hallucinations and out-of-body experiences. Many of those experiencing a NDE reported seeing bright lights. Bókkon, Mallick, and Tuszynski (2013) propose that these originate from the reperfusion of oxygen into the brain, resulting in the production of an excess of free radicals and other excited molecules that yield bioluminescence in the brain, which the brain interprets as come from an external source. Another possible explanation of NDE phenomena stems from the fact that different regions of the brain have different functions, so damage to different areas of the brain might account for the different experiences.

Evidence supporting the speculation that low levels of oxygen to the brain might be involved comes from studies of fighter pilots, who often experience a loss of oxygen to the brain from rapid accelerations (Whinnery 1997). They report sensations similar to those of people having NDEs, including out-of-body experiences, a pleasant sensation, memories of past events, tunnel vision, and bright lights.

One of the most interesting symptoms reported is an out-of-body experience. Patients felt that they were floating above the bed where they lay and could observe themselves and others in the room. Numerous tests have been carried out in an attempt to determine these experiences might be real or some sort of hallucination. For example, some experimenters have placed messages or objects on high shelves that would be above and out of sight of a patient in the bed; after the NDE, the patients were not able to remember seeing the hidden objects. Out-of-body experiences have also been reported by those with sleep paralysis; the cause of sleep paralysis is unknown, but it results in a feeling of being unable to more or talk. During an NDE, people may also have hallucinations, such as imagining something sitting on their chest. However, when questioned about their out-of-body experiences, they cannot accurately relate specific details of the room. For example, one person distinctly saw his body from above the bed, but the body he saw was wearing overalls, an item he never wore (Buzzi 2002).

Some part of the phenomena people report after NDEs may be explained by the "rubber hand" illusion, something that has been investigated in contexts unrelated to NDEs. In these studies the experimenter seats a subject in front of a fake hand. A screen hides the subject's own left hand from view. After the experimenter lightly strokes the fingers of the fake hand as the subject watches, and those of the real left hand behind the screen, for a few minutes, the subject comes to recognize the fake hand as the real hand. For example, when the subject is asked to use their other hand to point to their left hand, they point to the fake hand. To examine a person's awareness of their entire body rather than a single part such as a hand, Lenggenhager et al. (2007) used a variation on this illusion to show how subjects might have an out-of-body experience. They placed a camera behind a subject and had the subject wear a special headset that showed a virtual image of the person's own back about six feet in front of them. The subject's real back was stroked for a few moments, and they watched the virtual body's back being stroked. Then the blindfolded subject was moved away from where they had been standing, the blindfold was removed, and the person was asked to move back to where they had just been standing. The results showed that participants moved closer to where the virtual body would have been than to where they had been actually standing. The researchers interpret this to show that spatial unity and bodily self-consciousness are based on the processing of input from multiple senses, and they suggest that out-of-body experiences have a physiological explanation.

One might also wonder about the tendency of those reporting a NDE to fantasize. Martial et al. (2018) did just that. They used an established questionnaire (the Creative Experiences Questionnaire) to examine the fantasy-proneness of those who had had a NDE. All of the participants in the experiment also completed the Greyson scale about the NDE. Those who had reported a near-death-like experience (they feared for their life but were not actually in danger) scored higher on the fantasy-proneness survey than those who had actually had a NDE. The researchers concluded that fantasy-prone people might be more likely to report subjective experiences when prompted.

Blanke and Arzy (2005) extensively reviewed the literature on out-of-body experiences and concluded that the sensation of an out-of-body experience may be tied to a failure of the brain to integrate multiple streams of information being received from one's own body. This failure results in confusion about where the body is and leads those afflicted to believe their spirit has left their physical body. They conclude that all of the reported manifestations of these events can be explained easily by physiological actions. They also point out that a misfunctioning temporal-parietal junction may be involved in many of the experiences. The title of a paper by Mobbs and Watt (2011) sums up these conclusions: "There Is Nothing Paranormal About Near-Death Experiences."

From these reports, it is difficult to attribute any value to NDEs as a peek into the afterlife. However, they might illustrate some early aspects of the dying process as the hypoxic brain is struggling to survive. But in reality, the people who report NDEs have not come close to actually crossing the threshold to death, and thus we still know little if anything of what happens during dying.

Keeping Body and Soul Together

The question of what happens to a person's consciousness at death has been asked for hundreds of years. In 1907, Duncan MacDougall tried to measure the loss of the soul as a patient died. He carefully weighed dying patients just before and immediately after death. He reported that a person weighs ¾ ounce less after death, and he attributed that to the weight of the soul. While today science understands more about death, the question about consciousness has yet to be resolved.

Little is known about consciousness and dying because we know so little about how consciousness and the body are connected in the first place (Miller 2005). How is the physical brain of neurons and chemicals related to consciousness? During sleep, we lose consciousness to some extent, and every day thousands of patients are anesthetized and lose consciousness. General anesthetics affect specific areas of the cortex by blocking the brain's ability to integrate information (Alkire, Hudetz, and Tononi 2008). Wang, Eguchi et al. (2020) showed that a common anesthetic called isoflurane partially blocks the transmission of electrical signals at the synapse. However, it seems to selectively block higher-frequency impulses that control cognition and movement, rather than lower frequency ones that control breathing.

Other insights have been gained by studying the recovery of brain injury patients. Results suggest a connection between elements of physical recovery and the return of varying degrees of consciousness (Schiff 2010). Still, this does not take us too far toward understanding the precise elements of that connection.

The nature of consciousness may be the greatest problem left in biology. Today it lies in the realm of philosophy, but one might suspect that efforts in electrical engineering, computational science, biology, and physics may one day resolve these questions.

6

Ashes to Ashes and Dust to Dust

> In the sweat of thy face shalt thou eat bread, till thou return unto the
> ground; for out of it wast thou taken: for dust thou art, and unto dust
> shalt thou return.
>
> —Genesis 3:19 (King James Version)

In August 1984, workers discovered the body of a naked man in a peat bog in
northwest England. He was in remarkable condition, with his hair, his skin, and
some of his internal organs reasonably intact. But he was not a modern man.
He died somewhere between 2 BCE and 119 CE. When he died, he was about
twenty-five years old, sixty-six inches tall, and 130–140 pounds. He had a short
beard, good teeth, and fingernails that showed that he had not worked hard. For
his last meal, he ate bread made from wheat and barley. He seemed to have been
healthy at his death. He had some parasitic worms, but they did not cause his
death. His body had many wounds, but it was not clear whether these occurred
before or after his death. He appears to have been killed by a blow to the head or
possibly by strangling.

He came to be known as Lindow Man (Sammut and Craig 2019; Connolly
1985). Although his death was likely no accident, the preservation of his body
was. Ordinarily, his body would have decayed and been lost forever, but the
chemicals in the peat bog preserved or mummified him. He is now on permanent display at the British Museum.

Preserved bodies have been found in peat bogs in many areas, particularly in
Europe and Florida. Other bodies were found in the same bog as Lindow Man,
but they varied greatly in their state of preservation. Some were just a skeleton,
but others were well preserved, like Lindow Man. Some have been as old as
10,000 years, and others have been of Russian soldiers who died on the Eastern
Front in World War II.

The unusually good preservation of the bodies had to do with the conditions
in the bog—acidic water, low temperatures, and lack of oxygen. Interestingly,
the bodies were preserved in part by the death and decomposition of the plant
matter that makes up a peat bog. For example, humic or bog acid (see Figure
6.1) is a complex mixture of acids from decaying organic material (e.g., plants,

The Biology of Death. Gary C. Howard, Oxford University Press. © Oxford University Press 2021.
DOI: 10.1093/oso/9780190687724.003.0006

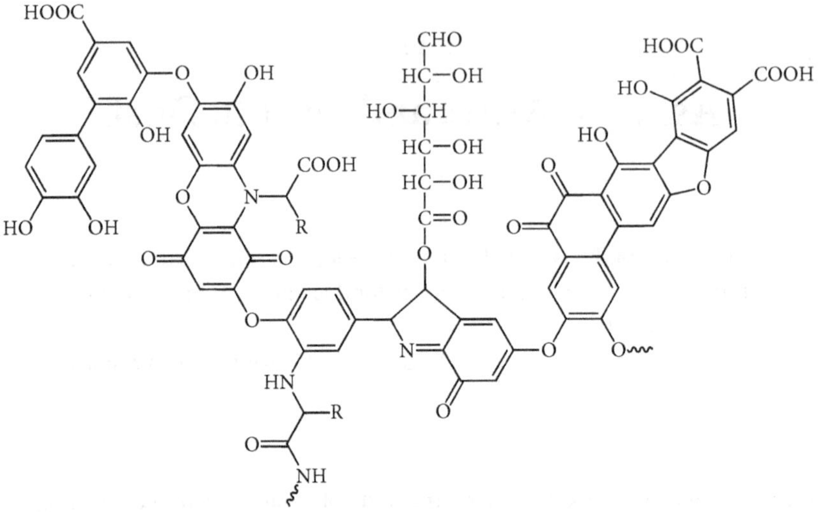

Figure 6.1 A typical humic acid found in a peat bog. The key functional groups include the phenolic (an oxygen atom connected by a double bond) and carboxylic groups (COOH). These groups interact with and cross-link collagen and other compounds in the body to help it resist decay.

Figure 6.2 Fossil fish. Herring (*Gosiutichthys parus*) from about 40 million years ago from the Green River Formation, Wyoming.

Photograph by author.

coal, soil). These acids "pickle" the bodies. Because bogs do not drain, the oxygen in the water is slowly used up by the chemical reactions of the decay processes. Without oxygen, the microorganisms that normally participate in decomposition cannot survive in the bog.

Peat bogs are not the only way that bodies can be spared from decomposition under unusual conditions. Other bodies have been well preserved in warm, dry sand or when frozen. Dried-out bodies decompose when exposed to water, and frozen bodies begin to decompose as soon as they thaw. Interestingly, babies who have never eaten lack internal bacteria; thus, their decomposition processes are different from those of adults, and they mummify quite easily under reasonably dry conditions.

Several frozen bodies of people who died long ago have been discovered. In 1991, a frozen man was found in the Alps, on the border between Italy and Switzerland (Bonani et al. 1994). He died a little more than 5,000 years ago. The reason for his death is not clear, but it was certainly violent: he had an arrowhead in his shoulder and had suffered a severe blow to the head. He, his clothes, and other artifacts were in remarkable condition.

As we've noted, these bodies were preserved under unusual circumstances. Most dead bodies undergo a more predictable—perhaps a more "normal"— decomposition. The normal process is accurately summarized by a phrase from the *Anglican Book of Common Prayer*: "Earth to earth, dust to dust, ashes to ashes."

What Happens After Death?

In this chapter, we undertake the delicate task of describing how animals and plants decompose after death. Our main focus will be humans, but the processes are quite similar in all animals, and they even share many common elements with the decomposition of plants. The natural processes might be challenging for some to read about, even a bit gruesome. If you are easily affected by this topic, you might want to skip to the next chapter.

Most of us know the basics of death. From a biological point of view, decomposition involves a fascinating array of reactions and myriad organisms. Once decomposition begins, it typically progresses at a reasonably predictable rate. This rate can be used to predict the approximate time of death, as many of us have learned from crime shows on television. Several factors influence this rate. As with all chemical reactions, it proceeds faster as the temperature increases. It also takes twice as long for bodies to decompose underwater as in the open air, and four times as long for those underground.

We have often sought ways to slow or even stop these inevitable reactions. We slow the process by embalming and burying bodies in coffins. As a result, some bodies are identifiable after many months. Usually after a year, only bones and teeth are left, but there are many examples of bodies that have remained fairly intact after some time. In 1901, the body of Abraham Lincoln was exhumed, so it could be reburied in a more secure tomb (Norton n.d.). To make sure they really had the actual body of Lincoln before they reburied it, they opened the casket. Even thirty-six years after his death, the body was said to be completely recognizable. In certain cases, conditions are just right to preserve structures to a greater or lesser extent, and more drastic methods can preserve the body for longer. The ancient Egyptians and others used mummification. Natural mummification and fossilization have preserved extinct animals and early humans. But in the end, we all return to dust.

However, unlike these unusual cases, most deaths follow a more predictable path. The cycles of birth and death form a natural rhythm. We are born and grow and live and die. The normal decomposition reactions recycle critical nutrients back into the soil, and these provide essential materials for other living organisms. In that way, decomposition provides a sort of "biological reincarnation"—a "rebirth" into millions of bacteria, insects, plants, and other living things that guarantees a life after death for our molecules and minerals and for the ongoing life of the planet. In fact, our molecules passed through many reincarnations before we took temporary possession of them. Thus, we only borrow the chemicals and elements for a while, and when life leaves our body, we return those chemicals and elements to the earth.

When the Heart Stops

We will leave the challenging question of when death actually occurs in Chapter 5. For the purposes of our discussion here, it is easiest to assume that death occurs when the heart stops beating or when blood no longer circulates. This assumption defines death in broad enough terms to include death of the entire organism, gangrene, necrosis of tumors, and other events.

Taphonomy is the term applied to the study of physical aspects of death. It was originally defined by I. A. Efremov (1940) as "the study of the transition (in all its details) of animal remains from the biosphere into the lithosphere." It is the study of how organisms—later expanded to include plants—that were once living decay and turn into fossils. Those events include what happens after death and before burial (biostratinomy) and what happens after burial (diagenesis). Taphonomy is described in five stages. Disarticulation occurs when the bones of the decaying animal are no longer held together by tendons. Like the tendons, the other non-skeletal parts of a dead organism are subject to breakdown and loss as part of the beginning of diagenesis. Dispersal occurs when those bones are scattered by various means, including water and other animals. Accumulation occurs when scattered bones collect in one place. Fossilization occurs when minerals seep into the organic material. Finally, mechanical alteration occurs when processes such as transport, burial, compaction, and freeze-thaw cycles change the remains. The five stages do not necessarily occur sequentially, and there is a great deal of interplay among them. Taphonomy is critical for understanding and interpreting fossils and other remains (Lyman 2010). For example, are the marks on a bone due to predation or to mechanical damage long after death? Taphonomy also has applications in forensics.

Once the oxygen that is supplied by fresh blood runs out, the tissue or the entire organism dies, and a relatively defined set of processes begins. Cells and tissues begin to break down in chemical processes called autolysis. (The term *autolysis* is a combination of the Greek words for "self" and "splitting"; it can also be thought of as self-digestion.) Without oxygen, an anaerobic environment is established. This environment favors bacteria normally found in the body, and they begin to break down the body's chemical components, including carbohydrates, proteins, and lipids (fats). A cascade of biochemical reactions begins, and the process accelerates as the cells break open. Enzymes are released from their normal cellular compartments where their actions are controlled and begin to react with other molecules that they would not normally encounter. Microorganisms that had been held in check by the immune system and other defenses of the living organism now multiply and spread. As the breakdown continues, various organic chemicals and gases are released. These reactions are well known, and they can be used to determine how long a person has been dead.

We will examine the sequence of events from three perspectives: the physiology, chemistry, and ecology of death.

Physiology

The first effects concern the whole body or its physiology (Parks 2011). As we noted in Chapter 5, first, when the heart stops beating and the circulation stops, the blood begins to settle to the lowest points in the body. The upper parts of the body become pale, and the lower parts become darker as the blood pools there. This is called postmortem lividity. It begins twenty minutes to three hours after death.

Second, the body begins to cool. Without circulation, tissues and their cells rapidly use their supply of oxygen, cease their energy-producing reactions, and begin to die. Some die quickly; others take longer. Brain cells die if deprived of oxygen for more than about three minutes, muscle cells survive for several hours, and bone and skin cells last for several days. In about twelve hours the body will feel cool to the touch, and in twenty-four hours it will cool to the core. After a couple of days, however, bacteria will grow within the body and cause its temperature to rise somewhat again.

Third, rigor mortis begins. Rigor mortis refers to the contraction of all of the muscles that sets in after death. It can make it very difficult to reposition the limbs. But what is it? To understand this process, we need to briefly review what muscles are and how they work. In the simplest terms, our muscles are made up of two kinds of fibers that are composed mainly of the proteins actin and myosin. When we contract a muscle, these two kinds of fibers "slide" past one another, shortening the length of the muscle. Another protein, troponin, acts as a bridge between the fibers, holding them in position. These contractions are controlled by calcium ions and other proteins. Normally, more calcium ions are found outside of the muscle cell than inside. When the muscle receives a signal to contract from our nervous system, pores in the muscle cells open and calcium ions rush in, causing the muscle fibers to slide past one another and contract the muscle. The muscle relaxes when the calcium ions are pumped back out of the cell in a process that requires energy. Without oxygen from circulating blood, muscles cannot produce the energy needed to pump the calcium ions out of the cell, so the muscles contract and remain in that state. Rigor mortis begins about three hours after death, usually in the jaw and neck, and lasts about twenty-four to thirty-six hours, until the actual muscle fibers themselves begin to decompose.

Chemistry

The next steps involve changes in the body chemistry, and the reaction times become longer. For the first few days, few signs are apparent, but a lot is going on. The general term for this process is *putrefaction*, and during this phase, the body changes dramatically. It is not a pretty sight.

The first stage is sometimes called green putrefaction. The intestines are affected early, and for good reason. Normally, the living gut is filled with bacteria. We need those bacteria to help with our digestion, and our systems keep them under control. However, after death, there is nothing to keep the bacteria at healthy levels. They grow rapidly and in an uncontrolled way, and begin to escape the regions of the body to which they are normally confined. They break down hemoglobin, the protein that normally carries oxygen in the blood, into sulfhemoglobin, and this turns the lower abdomen green. The bacteria also cause blood cells to break down, or hemolyze, which creates red streaks in the skin that soon turn green. Fluids form blisters on the skin and force it away from the flesh. Hair and skin come off the body easily.

A decaying body releases many chemicals and produces a putrid smell. Gases form in the abdomen, and the increasing pressure forces liquids and feces out of the body. Some of these chemicals have names that are surprisingly descriptive. For example, cadaverine and putrescine are the breakdown products of different amino acids. Within four to six days in temperate climates, or sooner in warmer environments, gases are released, and the body begins to swell and eventually rupture from the pressure. These gases include hydrogen sulfide (which smells like rotten eggs), methane, and traces of other sulfur-containing compounds, called mercaptans, that are related to those found in skunk spray. This phase exposes the soft tissues of the body, creating a feeding ground for scavengers and insects.

The next phase is black putrefaction. The skin turns black, bloated areas collapse, and fluid are released from the body.

As the tissues are lost and the bones are exposed, the body enters the mummification phase, when it dries out. Most of its bad odor disappears. The fats in the body begin to change. During life, we accumulate fats particularly in the abdomen, buttocks, breasts, and cheeks. About a month after death, they undergo chemical reactions that break them down into compounds called fatty acids and soaps. The term *soap* might be confusing for many, but it would be well understood by our ancestors, who rendered animal fat and mixed it with lye, usually from wood ashes; these basic materials formed a compound that yielded soap. In cadavers, this wax-like material is called adipocere, and it can help preserve the body because it resists bacteria, much in the same way as washing well with soap and hot water helps control the bacteria on our skin.

At this point, the decay process is in its final stage, which is called skeletonization. Most of the soft tissue is gone. Bones consist of both organic and inorganic material. The organic part is mostly a protein called collagen, which is the main protein in connective tissue. The inorganic part is a form of calcium phosphate called hydroxyl apatite. Even though bones are hard and strong, they still deteriorate over time. Bones are affected more by the acidity of soil and water than by bacteria or insects.

Ecology

We humans carry many organisms around with us during life, including a myriad of bacteria, fungi, and other microorganisms (Vass 2001). An adult human body contains about 37 trillion cells. However, these "fellow travelers" greatly outnumber us on a cell-per-cell basis, perhaps by as much as tenfold. Fortunately, most are benign. In fact, many are important to us. For example, they digest our food for us. In recent years, scientists have come to appreciate how critical these other organisms are to our own health. Of course, this applies only while we are alive and have an active immune system, intact skin that covers our body, and mucous membranes that line our body cavities. Those defenses keep the bacteria "out" of us as long as we are alive and producing energy.

With death, all of that begins to change, and our own ecosystem takes off in an uncontrolled manner. It may seem terrifying, but death provides a ready meal for many other living creatures. In fact, the recycling of nutrients is a critical process that provides those nutrients to other organisms. When the body dies, millions of microorganisms in the intestines survive; they attack the cells of the intestines and then move on to other body parts. The body itself adds to the decay by releasing its own enzymes and chemicals. For example, the pancreas has many enzymes that digest tissues. Cells have organelles called lysosomes and peroxisomes that break down proteins. As long as the organism is healthy, those organelles remain intact, and their enzymes are safely contained.

Organisms come and go in succession as specific food sources are consumed or exposed, and the invading organisms themselves will become food for yet other species (Hyde et al. 2015). In addition to bacteria and other microorganisms, lots of other creatures join in the banquet. Those organisms are fairly consistent. In fact, investigators can use the appearance of certain species to determine roughly the time of death. Insects, particularly flies and beetles, are attracted to the body by compounds released during decomposition. Scavengers move in and out of the body. Some feed on the body parts, and others feed on those initial grazers. Among the earliest are the blowflies (*Calliphoridae*) and flesh flies (*Sarcophagidae*). They lay eggs on the dead body usually within two days, but the timing can be quite variable. They favor areas around orifices (e.g., mouth, nose, ears, anus, penis, vagina, and eyes) or open wounds. Those eggs go through a series of stages with well-defined timing that allows fairly accurate estimates (e.g., egg, first instar larvae, second instar larvae, third instar larvae, prepupae, pupae within puparium, and imago). The pyralid moth (*Aglossa*) also typically appears early on, but some other organisms prefer bodies in considerable decay. The cheese skipper, *Piophila casei*, is a common pest that is often seen feeding on cheese and bacon; it invades the body three to six months after death.

Some species tend to be specialist scavengers that feed on particular tissues. For example, some beetles feed only on bone and thus will not be seen until later, when the bones have been exposed. Other beetles (e.g., rove beetles) feed on the blowfly larvae and appear only when those larvae are available. Beetles have chewing mouthparts and can attack the body before it has completely turned to liquid. Predatory beetles feed on fly larvae. They also lay their own eggs, which hatch into larvae that eat fly larvae. Beetle types that arrive later focus on specific niches, such as the skin or tendons. Carrion beetles (*Silphidae*) are more versatile in their diet. Some moth larvae (*Tineidae*) feed on hair. Parasitic wasps lay eggs inside the larvae or pupae of flies; the wasp larva hatches inside the maggot or pupa and feed on it until it dies, and then the adult wasp then flies off to repeat the cycle. Excretions of blowfly maggots contain a lot of ammonia; beetles in the genus *Necrophorus* find the ammonia toxic, so they can't coexist with the blowfly maggots. However, these beetles carry mites (*Poecilochirus*) on their bodies, and the mites eat the blowfly eggs, which reduces the population of blowfly maggots and thus allows the beetles to eat the carcass. Mites are tiny arachnids, more closely related to spiders, ticks, and scorpions than insects, which are more distant cousins. Humans also can carry several types of mites, such as gamasid mites that are usually involved early in decomposition, and tyroglyphid mites that arrive later to feed on the dry skin.

Factors that can confound the timing of decomposition include season, temperature, time of day, presence of water, and other factors. For example, knowing the specific insects in the area or recognizing dead insects in a body right after a spring thaw might help establish the date of death. Also, knowing the times when specific insects are available in an area can help.

Understanding the Process

The study of body decomposition has obvious applications in forensics. A great deal can be learned by examining a corpse, including cause of death, when and where the death occurred, and information about the dead person (such as race, sex, and age).

The Forensic Anthropology Center in the Department of Anthropology at the University of Tennessee has done outstanding research into the decomposition of human bodies. In 1980, they established a research facility—not open to the public—that follows the decomposition of donated remains under different conditions in a field setting. Their collection of more than 650 skeletons with a broad demographic span has been a valuable research and training tool.

By understanding the processes of decomposition, we have dispelled some common myths about the dead. For example, it is commonly believed that the hair and nails of the deceased continue to grow after death. This myth was well established long before Erich Maria Remarque powerfully described it in his classic novel *All Quiet on the Western Front* (1929):

> I cannot bear to look at his hands, they are like wax. Under the nails is the dirt of the trenches, it shows through blue-black like poison. It strikes me that these nails will continue to grow like lean fantastic cellar-plants long after Kemmerich

breathes no more. I see the picture before me. They twist themselves into cork-screws and grow and grow, and with them the hair on the decaying skull, just like grass in a good soil, just like grass, how can it be possible.

Despite Remarque's description, this is a myth. The hair and nails do not continue to grow after death. The skin does shrink somewhat, which exposes the roots of the nails and hair and can make them appear to have continued to grow.

Before bacteria and other natural phenomena were understood, the normal manifestations of human decomposition often struck terror in others, especially during epidemics. During the years of the Black Plague, death came often and rapidly. Mass graves were commonly used and reused. As graves were opened to make room for more victims, workers observed what seemed to be fat dead people who were full of blood. More shockingly, the death shroud that covered the body often had a hole at the mouth, and blood seemed to be coming from the mouth. As we now know, the "fat" person was actually full of gas from bacterial action in the decaying corpse, which would also push reddish fluid out of the mouth. The hole in the cloth shroud was caused by bacteria from the mouth that ate the cloth. The workers were so scared by these "vampires" that they would try to "kill" them; a stone or brick might be forced into their mouths to prevent them from feeding on living humans.

In other situations, understanding the processes of death may allow us to put them to use in some ways. Meat, particularly wild game, has long been hung to age, a practice that began well before the underlying mechanisms of decomposition were understood scientifically. In fact, the very best beef are still hung to dry for several weeks to develop the best taste. The flavor increases dramatically because moisture evaporates from the meat and enzymes in the meat break down the connective tissues (collagen), resulting in much more tender meat. Fungi from the genus *Thamnidium*, among others, can form a crust on the outside of the meat. It is trimmed off before cooking, but the enzymes from the fungus add to those from the meat itself to tenderize and flavor it. The process of aging takes fifteen to twenty-eight days, and over the course of that period about one-third of the weight of the meat is lost. Typically hanging is reserved only for unusually high-quality and large cuts of meat with a good amount of fat evenly distributed throughout.

Laying the Dead to Rest

One of the features of human civilization is the recognition of death and the ceremonial burial of the dead. Both early and modern humans did not simply walk away from their family members and friends when they died. They showed great respect for them and treated them tenderly. The dead were buried or placed in a cave or other location in what seemed a ritual manner. Often decorative objects or materials were included, things that people thought would be needed by the dead person in the afterlife. Across a variety of cultures through time, corpses have been "washed, embalmed, anointed, pickled, dismantled, painted, adorned with jewelry, clothed, wrapped, placed in a container, moved, viewed extensively, touched, embraced, wept over, shouted at, danced over, and force-fed food, among other practices" (White, Marin, and Fessler 2017). Burial sites from 300,000–500,000 years ago have been found, but not everyone agrees that they were burial sites. Neandertals were thought to intentionally bury their dead; by 100,000 years ago, modern humans were clearly burying their dead.

In 2013, two men were exploring the Rising Star cave system in South Africa. The caves had been known at least since the 1980s, but there are many passages within the cave system. They found a very narrow, nearly vertical passage that led to a large room, now called the Dinaledi Chamber. That part of the cave was nearly inaccessible and totally dark. However, the floor of the room was littered with skeletal remains. The fifteen skeletons were the remains of individuals from an archaic hominid species called *Homo naledi*. The remains seem to have been carried there intact; there were no indicators of predators. Since then, the remains have been dated to between 236,000 and 335,000 years ago, making this the earliest hominoid burial site so far found (Dirks et al. 2015, 2017).

Some cultures have taken treatment of the dead to the next step by trying to preserve bodies after death, and perhaps the most famous of these were the ancient Egyptians. Many of their mummies are in surprisingly good condition, even after thousands of years.

Mummification was an elaborate procedure and so was reserved for royalty and other important people (Abdel-Maksouda and El-Aminb 2011). First the dead body was washed with palm wine. Then an incision was made in the left side of the torso to remove the internal organs, except the heart. The heart was considered the center of the individual and too important to be removed. The other organs, such as the liver, lungs, stomach, and intestines, were washed and packed in natron to dry and preserve them. Natron is a naturally occurring mineral. It is a mixture of salts, including mostly sodium carbonate decahydrate and about 17 percent sodium bicarbonate (baking soda). A long hook was pushed into the nose and into the skull to pull out the brain. Then the body was packed with and covered in natron. After forty days, the body was washed with water and the skin was treated with oils. Early on, the organs were placed into canopic jars, but in later times, the dehydrated organs were wrapped in linen and returned to the body. The body was then packed with sawdust, linen, and leaves to fill it up so it would look natural. Finally, the body was carefully wrapped in linen. The head was wrapped first. Then the fingers and toes were each wrapped separately. Finally, the whole body was covered with linen. At each step, the linen was painted with resin to hold the pieces of cloth together.

Today we also attempt to preserve the dead. We call it embalming, and Ambrose Bierce (1911) humorously described it as follows.

> EMBALM, v.i. To cheat vegetation by locking up the gases upon which it feeds. By embalming their dead and thereby deranging the natural balance between animal and vegetable life, the Egyptians made their once fertile and populous country barren and incapable of supporting more than a meagre crew. The modern metallic burial casket is a step in the same direction, and many a dead man who ought now to be ornamenting his neighbor's lawn as a tree, or enriching his table as a bunch of radishes, is doomed to a long inutility. We shall get him after awhile if we are spared, but in the meantime the violet and rose are languishing for a nibble at his *glutoeus maximus*.

The dead in the United States and a number of other countries are normally embalmed before burial. The procedure is not so involved as those of the Egyptians, and the results are far less permanent. All the clothing is removed, and the body is washed carefully with disinfectant. The blood is removed through major blood vessels in the neck, usually by injecting embalming fluid into the right common carotid artery and draining the blood from the right jugular vein. Other body fluids are removed and replaced by embalming chemicals. A small incision near the navel, and a trocar is used to empty the body cavities (e.g., stomach, bladder, intestines) and to inject embalming fluid. In some cases, embalming chemicals are injected under the skin if those areas will be seen during a funeral service. Several chemicals might be used. The major one is formaldehyde. It cross-links proteins so that they are less likely to be broken down. Creams are used to keep the skin soft. Dyes make the skin color more natural. Even embalmed bodies decay, although at a slower rate.

Other Methods

Cremation

Cremation of human bodies has been practiced for thousands of years. The first known cremation occurred 17,000 years ago at Lake Mungo, Australia (Gillespie 1997). The practice has been in and out of favor with different groups at different times throughout history. Currently, cremation is the method of choice for most of the world. The United States is catching up; already about half of the US dead are cremated. The National Funeral Directors Association predicts that by 2035, only 15 percent of interments in the United States will involve a traditional burial. Cremation is cheaper and more efficient, especially for a population that increasingly has little connection with the land. In modern times, cremation has been viewed as an environmentally responsible. There are no caskets, headstones, or toxic chemicals, and no land is used for burials.

Bodies can be cremated on open funeral pyres or in a crematorium. In the United States, bodies are cremated in furnaces that generate temperatures of 1,600–1,800°F. The time required varies but is generally about one hour for every 100 pounds. Organs and soft tissues are vaporized and oxidized. The gases are released into the outside air. All that remains are bone fragments, which are usually ground to a powder about the consistency of sand. The "cremains," as they are called, usually amount to roughly 3.5 percent of the original body weight, about 2.5 kilograms for an average person.

Liquid Cremation

Liquid cremation involves the alkaline hydrolysis of the body (Olson 2014). The body is placed in a vessel that can be pressurized. Potassium or sodium hydroxide, very strong bases, is added, and the vessel is put under pressure and heated to 320°F. The high pressure prevents the mixture from boiling, but it speeds the chemical reactions that chemically break down the body. After four to six hours, the body is reduced to very soft bones that are ground to a fine powder and a brownish green liquid containing peptides, amino acids, sugars, salts, and soap.

Proponents of liquid cremation point to the facts that it uses less energy, neutralizes infectious agents and embalming chemicals, and avoids the use of cemetery space. Others see the practice as disrespectful, akin to flushing the deceased down the drain. The legality and availability vary from state to state.

Human Composting

To many people, the options available for dealing with their dead loved ones seem to be undesirable. Rather than be filled with preservatives or burned, they would rather contribute to the living world more directly. A new method called recomposition or human composting offers that possibility and, thus, has considerable appeal as an ecofriendly alternative to traditional burials and cremation. The process is very similar to natural decomposition, but it is speeded up by carefully controlling the breakdown reactions. As in traditional composting, the remains are allowed to decay by normal processes. The process takes place in a container. The body is covered with wood chips, alfalfa, and straw, and the levels of oxygen and water are controlled to accelerate the transformation. The reaction temperature rises to 120–160°F. Bacteria and other microorganisms do the rest within about a month. Even teeth and bones are transformed. A human body yields approximately one cubic yard of compost. In March 2020, the practice became legal in Washington state (Solly 2019).

Other "green" burials have come into use in some places. For example, some companies (e.g., Coeio) offer a "mushroom suit" that contains mycelium, the threadlike structures of mushrooms that live underground. They accelerate the decomposition of the body after burial. Other types of burial pods contain the body and a tree is planted over the pod. The tree then feeds off the decomposing body.

Animals and Death

Animals and humans perceive death differently, and while we now better recognize some emotional behaviors in animals, there is no known example of animals performing any rituals after death. The concept of death usually has four features. First, all living things eventually die. Second, death cannot be reversed. Third, the dead cannot think, feel, or do anything else. Fourth, death results from something going wrong in the organism. Even human children have trouble understanding these features. Many animals remove dead individuals from their next or burrow (Anderson 2016). Special undertaker bees remove dead bees from the hive and fly them away for disposal. Among bees, dead individuals are noticed because of the fatty acids that they release, particularly oleic acid. In fact, a dot of oleic acid on a healthy individual will result in that individual being removed from the hive by the undertakers.

Legend of Squanto and Fish as Fertilizer

Life and death are part of a larger cycle that ensures that life will continue. We learn at an early age that decomposing organisms provide nutrition for plants. Most elementary school students in the United States learn the story of how Squanto taught the Pilgrims to use a fish to fertilize their cornfields and saved them from starvation. The story may be true, but using whole fish as fertilizer was not a farming practice used by Native Americans. Squanto's life involved an amazing odyssey that took him as a slave from New England to Spain and then to England and on several voyages of discovery before returning him to his original home. Modern historians believe Squanto probably learned the practice of fertilizing crops in Europe.

Nevertheless, dead fish are an excellent fertilizer. In fact, as they die after spawning, Pacific salmon (*Oncorhynchus* spp.) provide a significant source of nutrients in rivers. Each year the deposition of nutrients derived from the ocean and borne upstream by salmon are important for the many organisms in freshwater communities throughout the Pacific coastal region. Many food webs in streams benefit. Other fish eat the dead salmon. Periphyton—a name for a variety of organisms, including algae, cyanobacteria, and heterotrophic microbes, mostly attached to surfaces underwater—is an important source of food for aquatic invertebrates, tadpoles, and fish. Terrestrial vertebrates also scavenge the dead salmon, and their nutrients benefit riparian soils and plants as well.

Do animals ever recognize death or experience grieve over death? For a long time, the answer was generally considered to be no. The view of animal recognition of death is reflected by a statement by Ernest Becker in his book *Denial of Death* (1973, 51). He asserted that animals know nothing about dying and death: "The knowledge of death is reflective and conceptual, and animals are spared it."

Since that time, biologists have taken a more expansive view of the emotional capacity of animals. More recent reports document some form of recognition of death in nonhuman mammals. Most concern the behavior of mothers regarding dead offspring. Often mothers carry the dead baby for varying lengths of time. They may also participate in other behaviors, such as grooming and protecting the dead infant. These behaviors are seen in a variety of species, including apes, monkeys, dolphins, dingoes, and giraffes. Unfortunately, many of these reports are anecdotal (Watson and Matsuzawa 2018).

Some animals have also been seen to care for dying group members. In a case study (Yang, Anderson, and Li 2016), wild Sichuan snub-nosed monkeys (*Rhinopithecus roxellana*) were observed as they tended for a dying adult female. Members of the group and the lead male groomed her and touched her hand. Once she died, the others stayed with her for a while, and the male stayed longer and repeatedly touched her. The next day they returned to the place where she had died. These actions are similar to those observed elsewhere in other great apes.

Chimpanzees have been seen using tools to clean the body of a dead member of their group (van Leeuwen, Cronin, and Haun 2017). After a young male died, a female was seen using a grass stem to clean his teeth. This is anecdotal evidence, to be sure; however, observations in nonhuman mammals are important.

They may help us to gain a better understanding of how human emotions around death have evolved. Additional studies of animal behavior and of early human burial sites will help to resolve this issue.

How Plants Decompose

Like humans and animals, plants die and decompose. Many of the processes are similar, but considerably less gruesome. The breakdown of plant material is critical to recycle nutrients to feed new plants that animals and humans ultimately depend on.

While plants are living and growing, they use photosynthesis to capture carbon dioxide from the air and convert it into sugars and starches that later form energy sources for animals and humans. Plants also synthesize vitamins and other important nutrients or otherwise make them available. For example, nitrogen is a critical element for life, and although the air all around us is 80 percent nitrogen, animals cannot capture it. Some types of organisms, such as blue-green algae, lichens, free-living soil bacteria, and bacteria that live in the root nodules of certain types of plants, convert nitrogen gas to useful forms (nitrites and nitrates) in a process called nitrogen fixation; the fixed nitrogen stays in the plant. When plants are plowed under or allowed to decay in the field, the nitrogen they contain becomes available for other plants. Plants also feed on the nitrogen-containing compounds from dead bacteria.

As with animal decomposition, an ecosystem forms when plants begin to decompose. That ecosystem contains organisms that feed on various parts of the plant. They all take their turn eating as different molecules are exposed in different parts of the process. Primary consumers, including bacteria, fungi, nematodes, some types of mites, snails, slugs, earthworms, millipedes, and sowbugs, eat the organic matter. Secondary consumers, such as springtails, other types of mites, some types of beetles, nematodes, rotifers, and soil flatworms, eat the primary consumers. Tertiary consumers, such as centipedes, ants, spiders, and other types of beetles, eat the primary consumers. The higher-level consumers contribute excrement and their own dead bodies to the organic material to begin the process again.

Earthworms participate in decomposition in two ways. They eat dead plant material and leave their droppings. In addition, the holes they dig allow oxygen and water to penetrate the decaying material, giving other small decomposers more access to the material and encouraging chemical reactions.

Also in soil, the decomposed cell walls of plants remain and help bind together clay particles and other elements of soil. This gives the soil pores so that air and

Figure 6.3 Rotting tree stump. Plants decompose in a series of biochemical and microbial reactions that are critical for recycling key nutrients in the environment. Photograph by author.

water can penetrate into the ground, and it gives a structure for the organic components to stick to so they are not washed away.

So what exactly happens during plant decomposition? While many of the degradation processes are similar for plants and animals, there is one significant difference. The cells of both plants and animals have a cell membrane, but plant cells also have a tough cell wall made of cellulose, hemicelluloses, and lignin. The first step in plant decomposition is breaking the cell wall. Of course, in biology, where there is a need, there is a solution. Many species have evolved to take part in the breakdown, including invertebrates and insects in the soil and fungi. These materials of the cell wall decompose mostly by biochemical methods and at different rates, depending on temperature, soil conditions, and water. Lignin, for example, resists most organisms, except some fungi, such as white rot fungus.

The transformation of organic matter is a fascinating process. Fungi and bacteria start the process. They have enzymes that break down cellulose and other plant components (enzymes that many small animals lack). The biochemical compounds that make up plants are extremely complex, and the organisms involved in decomposition use a variety of enzymatic reactions to decompose plant material. Oxidation is an important reaction. This is the same basic reaction that occurs when wood burns, but the reaction is much slower. A plant carbohydrate, protein, or lipid combines with oxygen to produce carbon dioxide and water. In biochemistry, the carbohydrate or other compound is oxidized and the oxygen is reduced. Thus, these reactions are called redox (for reduction and oxidation) reactions.

Other enzymes attack specific molecules. Chitinases catalyze, or speed up, the degradation of chitin. They have been found in many organisms, such as bacteria, fungi, plants, invertebrates, and vertebrates. Endochitinases cut chitin inside the molecule, and exochitinases degrade chitin from the ends. Cellulases break down cellulose, a complex carbohydrate. Various proteases digest proteins either by working from the ends of the protein molecule or by cutting in the middle.

The most important microorganisms are bacteria, and they make up about 80–90 percent of the microorganisms in the soil and in decomposing matter. Actinomycetes give soil its "earthy" smell. In decaying material, they look like spiderwebs with long gray filaments. They degrade complex organics (cellulose, lignin, chitin, and proteins). Fungi (molds and yeasts), which tend to live on the outside of the decay, also break down complex plant components and, in that way, help bacteria to do their work. Protozoa (one-celled animals), rotifers, and other microscopic animals that live in water droplets feed on the organic material, bacteria, and fungi.

The decomposition reactions are greatly affected by the reaction conditions. As just noted, oxygen is used in the reactions, and carbon dioxide is produced. If all of the oxygen is used up, a different set of reactions will take over. These use sulfur and produce a smell like rotten eggs. Acidity is also critical. The pH scale, which varies from 1 to 14, is used to measure acidity. A relatively neutral pH of 5.5–8.5 is ideal for decomposition. The reactions themselves release organic acids, which early on, favors the growth of fungi.

Composting is a process of aerobic decomposition of plant matter that many people are familiar with. Many kinds of microorganisms are involved, and each

prefers a particular type of organic material. Composting also depends on physical conditions, such as temperature, moisture, and aeration. Decomposition
reactions release heat, and the temperate of a compost pile can reach 40–50°C
(104–122°F) within a few days. The heat encourages growth of microorganisms
that are involved in the decomposition. But there is a limit: if the pile gets too hot,
the microorganisms will die. This is why the pile must be turned periodically. It
aerates the pile and ensures that the reactions continue optimally. The amount of
water in the pile affects the temperature and the reactions, as well, as water tends
to hold heat. If the plant material has been broken down too much, the small
particles can become compacted and block the circulation of oxygen, which will
slow the reactions.

Sometimes plants (like animals) do not decompose, becoming fossils. The
process of fossilization can involve either permineralization or replacement. In
permineralization, or petrification, water containing large amounts of calcium
carbonate or silica is absorbed by relatively porous organic plant material. Then,
time and pressure stabilize the minerals and preserve the plant structures. In replacement, water dissolves the plant material, and the remaining space is filled
with the minerals. The result in either case is a fossil replica of the original living
tissue.

Plants can be preserved over the long term in a different manner as well. In
the Carboniferous period (300–350 million years ago), tree ferns (*Filicales*),
horsetails (*Equisetales*), club mosses (*Lycopodiales*), and other plants covered
much of the earth in swampy forests. When those plants died, the water and mud
of the swamps hindered normal degradation. Some of the dead organic material
decayed under anaerobic conditions (in which little or no oxygen was present),
forming a thick layer of peat. In some cases, millions of years' worth of heat and

(a)   (b)

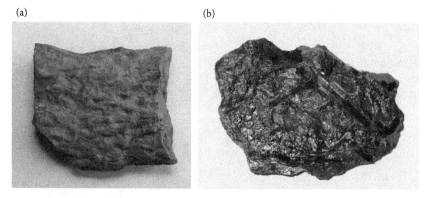

Figure 6.4 Fossil plants. (a) Petrified wood. (b) Fossil in coal.
Photographs by author.

pressure from overlying sediment squeeze the water out of peat, changing into it coal. Coal, sometimes called "fossilized sunlight," is more than 50 percent carbon by weight. There are several types of coal, including lignite, sub-bituminous coal, bituminous coal, anthracite coal, and graphite (pure carbon). Ten feet of peat produce about one foot of coal.

Like coal, petroleum also comes from living sources. Zooplankton and algae died and settled to the bottom of a sea or lake. They later were covered by mud and other sediments. After millions of years of heat and pressure, they changed into various compounds. For example, kerogen is organic material found in oil shale. Kerogen contains carbon disulfide or bitumen, which the ancient Egyptians used to prepare mummies of their dead. It is also useful for waterproofing boats. More heat and pressure turn kerogen into oil and natural gas.

As Carl Sagan (1997) noted in his book *Billions and Billions*, "Coal, oil and gas are called fossil fuels, because they are mostly made of the fossil remains of beings from long ago. . . . Like some ghastly cannibal cult, we subsist on the dead bodies of our ancestors and distant relatives."

7

Genetics of Life and Death

> Maybe one day we will be able to take a pill that keeps us young and
> healthy much longer. I believe in my heart that this will happen.
> —Cynthia Kenyon (quoted in Cook 2005)

Could Dr. Kenyon be right? Could we take a pill and live longer, or even avoid
death? Can we slow aging or even stop it? Humankind has long hoped that the
answer might be yes. In some way, we all hope we can cheat death to live longer.

It's not a new dream. According to legend, the Spanish explorer Juan Ponce
de León landed in Florida in 1513 seeking the Fountain of Youth. Historians
believe that Ponce de León was most likely looking for gold and other riches,
and that the legend of the Fountain of Youth was invented years later and preserved to promote tourism. Still, Ponce de León would not have been the first
to seek eternal life. Humankind has been looking for immortality for centuries.
The ancient Chinese believed that ingesting gold, jade, and other precious minerals would give them some degree of immortality; accounts by Herodotus, the
legends of Alexander, and stories from Polynesia also refer to immorality. In the
Epic of Gilgamesh, an ancient poem from Mesopotamia, a man named Enkidu is
killed, and his friend Gilgamesh spends the rest of his life and the last part of the
poem looking for the secret of eternal life.

Although most people accept these stories as only myth, Americans and
others continue to seek ways to avoid aging and death. In 2018, more than
17.7 million cosmetic surgical and nonsurgical procedures were performed in
the United States, according to the American Society for Plastic Surgery (ASPS
2018). In 2016, the total spent on all cosmetic procedures was more than $16 bil-
lion (ASPS 2017). Several of the most common cosmetic procedures are aimed
at the effects of aging, such as forehead lifts, dermabrasions, breast lifts, face lifts,
tummy tucks, liposuction, and eyelid surgeries.

Many more billions of dollars are spent on cosmetic products in search of
youth every year. In fact, the total worldwide market for cosmetic products, including anti-wrinkle creams, hair color, and hair-restoration treatments, is estimated to be more than $500 billion (Statista 2020). Thus, even without Ponce de
León, the search for the Fountain of Youth continues.

The Biology of Death. Gary C. Howard, Oxford University Press. © Oxford University Press 2021.
DOI: 10.1093/oso/9780190687724.003.0007

The average life span in the developed world has increased significantly over the last century. In 1900, most Americans did not live beyond age fifty. Now Americans can expect to live well into their eighties. The number of Americans eighty-five years old and over is expected to nearly double from 2016 to 2035 (from 6.5 million to 11.8 million) and nearly triple by 2060 (to 19 million people) (Vespa, Medina, and Armstrong 2020). Many of us know people who are 100 or older. At any given point the record holders for the world's oldest human typically range from 112 to 115 years; the oldest known person ever was Jeanne Louise Calment of France, who lived to be 122 years and 164 days old (Robine et al. 2019). Clearly, humans—at least some of us—can live more than 100 years.

As noted in Chapter 4, at first glance, it might seem reasonable for people to die at some point after they reproduce. The death of older individuals might help conserve scarce resources and thus be advantageous to those of reproductive age. However, as discussed earlier, there is no way to square this idea with natural selection. Genes that benefit the fitness of individuals before and during the reproductive years are subject to selection. Events that occur after the reproductive period are not subject to selection. Thus, genes that have their effect later in life cannot be selected for or against.

Recently, however, molecular geneticists found intriguing hints that we might be able to live longer, if not eternally. They suggest that many diseases, such as heart disease, diabetes, cancer, and brain disorders, may just be symptoms of aging. Aging itself might result from many factors, including damage to our bodies, organs, and cells that adds up over many years, and the genes we inherit from our parents. It's been discovered that different organisms seem to share some hallmarks of aging, such as genomic instability, telomere shortening, changes to epigenetics, loss of proteostasis, cell senescence, dysfunctional mitochondria, and more (López-Otín et al. 2013). In the following pages, we will look at the attempts to identify some interesting genetic targets and review some of the most promising areas.

Theories of Aging Based on Genes and DNA

Genetics is often a good place to begin thinking about biological problems. Many people survive into their nineties. Moreover, in some families people seem to live longer than in others, and we might hear comments such as "Long life runs in their family" or "They have good genes."

Could that be true? Are there genes that allow people to live longer or shorter lives? Maybe, but how could we explore that? One way is to look at the extremes of aging in humans.

"Good" Genes

In a famous anecdote, Willie Sutton was asked why he robbed banks. His answer: that was where the money was. So if there are genes that allow for long life, one might expect to find them by comparing the genome of people who have lived a long time to those of people with more ordinary life spans. Twin studies have already suggested that genetics account for 20–30 percent of the differences in human life span. Another factor worth investigating is that centenarians not only live longer than others but also seem to have better health for longer than others (Hitt et al. 1999). Somehow they elude many of the normal age-associated diseases, such as cardiovascular disease, cancer, stroke, and diabetes.

The New England Centenarian Study (https://www.bumc.bu.edu/centenarian/) is looking for those genes. They collect blood samples and medical histories from centenarians around the world. With these samples, researchers might find ways to extend human life spans. The study has identified several families with amazing longevity. For example, in one family, six of ten siblings had lived longer than 100 years.

After studying the genomes of these older people, researchers noted an intriguing link to a region of chromosome 4. That was the good news. The challenge was that the region is 12 million base pairs long; that's enough DNA for hundreds of genes, so looking for the relevant genetic information within that area is not an easy task. Amazingly, they found a single-nucleotide polymorphism (SNP, a substitution of a single nucleotide at a particular spot in a gene) in the gene for microsomal transfer protein, which is involved in cholesterol metabolism (Puca et al. 2001). But that was only one gene, and while in lower organisms with much smaller genomes, such as nematodes, fruit flies, and yeast, only a small number of genes, perhaps half a dozen, affect life span, human longevity might be a complicated trait that involves multiple factors. The Human Genome Project and rapidly improving tools may help here; for example, with gene chip technology, researchers can study thousands of SNPs or mutations at the same time, in a single experiment.

Fortney et al. (2015) statistically examined genome-wide association studies of Alzheimer's disease and coronary artery disease to look for connections with life span. More specifically, they looked for SNPs associated with longevity. They found eight SNPs and were able to link four of them to factors affecting diseases normally associated with aging. These four genes involved apolipoprotein E (associated with Alzheimer's disease), CDKN2B (associated with cellular senescence), ABO blood groups (centenarians are more likely to have the O blood group), and SH2B3 (associated with extended life span in fruit flies and involved in neurological disease in humans).

Other gene-association studies of long-lived individuals have focused interest on different polymorphisms in a few genes. For example, the genes for insulin-like growth factor 1 receptor gene (IGF1R) and the FOXO3 transcription factor are associated with long life (Pawlinkowska et al., 2009). Levels of apolipoprotein E4 are reduced; as we saw in Chapter 3, apoE4 is associated with a higher risk of Alzheimer's disease and problems with cholesterol metabolism.

However, Gierman et al. (2014) completed whole-genome sequencing of seventeen individuals who were 110 years or older. They did not find any rare protein-altering variants associated with longevity. So while genetics is clearly important, how it is so is not clear at all.

Born Old

While some researchers have been looking at individuals who live exceptionally long lives, others have sought clues from those who age much more rapidly. As has been mentioned briefly in previous chapters, aging occurs quite rapidly in a very small group of children with a disorder called progeria or Hutchinson-Gilford progeria syndrome (HGPS) (Cao and Hegele 2003; Merideth et al. 2008). It is very rare, occurring in about one in every 4 million births, according to the National Institutes of Health. Children with this genetic disease seem normal at birth, but within six to twelve months, they begin to show signs of aging, including changes to their skin and loss of hair. They grow slowly and do not gain weight at the normal rate. Their faces feature prominent eyes, thin noses, small chins, and protruding ears. Tragically, their average life expectancy is only about thirteen years. Most die from atherosclerosis, or hardening of the arteries, affecting the heart and brain.

The cause of HGPS is known to be a mutation affecting the gene that encodes the protein lamin A. Lamin A forms part of the scaffold supporting the membrane around the cell's nucleus, and so is critically involved in maintaining the structure and organization of the nucleus. In HGPS, the diseased nuclei are misshapen, which some researchers believe renders cells unstable. Interestingly, in laboratory experiments Scaffidi and Misteli (2005) were able to reverse the cellular effects just mentioned by correcting an aberrant splicing event. The authors believe that these results strongly indicate that small-molecule drugs are likely to cure HGPS at some point in the future.

Although HGPS is caused by a mutation in a gene, it is not inherited. The mutation occurs by chance. Nevertheless, there is some evidence that other progeria-like syndromes run in families. In Wiedemann-Rautenstrauch syndrome, aging occurs before birth. In Werner syndrome, aging begins later in adolescence or early adulthood. In these two diseases as well, the rapid aging leads to early death.

More evidence about the mechanisms involved in HGPS came from research by Stephen G. Young at UCLA, who studies isoprenylation, a specific type of change to a protein (Yiping et al. 2016). His team used gene-targeted mice to study the enzymes that oversee a complicated processing reaction in some proteins, including the lamins. His team knocked out the gene *ZMPSTE24*, which encodes an enzyme in this pathway called endoprotease, and found that these mice could no longer make mature lamin A, but rather only a premature form called prelamin A. The *ZMPSTE24*-deficient mice have many symptoms of aging that resemble those in humans with progeria. The specific mechanism seems to be that prelamin A is toxic. Reducing its levels by 50 percent eliminates the misshapen nuclei in cultured cells and completely "cures" the disease symptoms in the *ZMPSTE24*-deficient mice. They also found that blocking a reaction involving prelamin A lessens the aging-like signs in *ZMPSTE24*-deficient mice. These studies suggest possible ways of treating children with progeria.

Epigenetics

As we've seen, epigenetics—the way genes are expressed—influence many aspects of development, and it's not surprising that some scientists have hypothesized that it is involved in the regulation of aging. In an extensive review of animal-, tissue-, and cell-based studies of aging, Sen et al. (2016) found links between aging and epigenetic changes that affect the structure of a DNA-protein complex called chromatin. Other epigenetic factors may also affect histones (proteins that bind tightly to DNA), the positions of nucleosomes (particular combinations of DNA and histones that are put together to make chromatin), and regulation of telomere length (Song and Johnson 2018). Epigenetic treatments have two significant advantages: they are reversible, in that withdrawing the treatment returns the expression of the gene to the way it was before treatment was started, and they are relatively stable. Stability means that the organism can put genetic programs in place that not subject to easy change. One drawback is that such treatments may control large segments of the genome, and so they must be used with great care.

Ocampo et al. (2016) used four specific transcription factors (the Yamanaka factors—Oct2, Sox2, Klf4, and c-Myc [Liu et al. 2008]) to partially reprogram cells in a mouse model of aging. They found that the typical hallmarks of aging were reversed and the life span of the mice was lengthened. These findings support the concept that problems with epigenetic regulation are a driver of aging.

Nutrient-Sensing Mechanisms

Eat Less and Live Longer?

Surprisingly, in all organisms tested so far, life span can be increased by one simple method: limiting food intake. This effect of calorie restriction was first noted in yeast, but it also works in the worm *Caenorhabditis elegans* and the fruit fly *Drosophila melanogaster* (Bordone and Guarente 2005). Most interestingly, a similar effect has been seen in higher organisms. Mice consistently fed 30 percent fewer calories live about 30 percent longer than normal. Some short-term studies in humans and longer-term studies in monkeys suggest that calorie restriction may be of benefit to us too.

Sounds great—eat less and live longer. Unfortunately for us, it isn't that easy. To have any effect in humans, caloric intake must be extremely restricted. To even moderately extend life span, a human could only consume 1,000-1,200 calories per day forever. This level is near to starvation, so it's clearly not recommended. However, short periods of extreme calorie restriction, such as a day of fasting, have shown benefit. In addition, many people have practiced "time-restricted" feeding or intermittent fasting, which involves eating only during an eight- or ten-hour period each day. That strategy has some support. Researchers working with mice discovered that confining caloric consumption to an eight- to twelve-hour period might stave off high cholesterol, diabetes, and obesity (Longo and Panda 2016). They are exploring whether the benefits of time-restricted eating might apply to humans as well as mice. A related topic for investigation is whether long bouts of exercise, which force the body to rely on fat for fuel instead of carbohydrates, may also extend life span.

If we could just learn how caloric restriction works to help increase life span, we might be able to develop drugs that would yield the same benefit. To that end, researchers have looked at the genetic basis of cell signaling pathways (that is, the way groups of molecules work within a cell to control various types of cell function). One target for investigation is a molecule called insulin/insulin-like growth factor (IGF-1), which operates in the IIS signaling pathway (Tatar, Bartke, and Antebi 2003)—a signaling pathway that emerged early in the evolution of living things on earth, as far back as the appearance of multicellular organisms. Mice with mutations in the genes that affect this pathway have improved glucose homeostasis, immune function, and neuromuscular performance, but how these affect longevity is unclear.

A team of scientists at the Massachusetts Institute of Technology, led by Leonard Guarente, discovered in yeast a gene, *SIR2*, that they propose plays a role in the longevity effects of caloric restriction (Guarente 2007). In some way, sirtuin, the protein encoded by this gene, "senses" that the organism is running out of nutrients and makes many adjustments to the metabolism to help the organism survive. Though, as we've seen in Chapter 4, "aging" in yeast is not exactly the same as in humans (researchers define yeast life span in terms of number of cell divisions), most animals have genes that are analogous to *SIR2*. In mice and humans, there are seven genes, *SIRT1* through *SIRT7*. *SIRT2* is similar to the yeast gene *SIR2*, and the other proteins they encode have related functions. The proteins are enzymes that modify histones, proteins that bind tightly to DNA and help organize our genes, and other proteins. These modifications (which come about through chemical processes such as phosphorylation, methylation, or acetylation) can change the shape or conformation of the protein, turn on or off its function, or change its structure.

The functions of many of the sirtuins implicate them in aging. They are directly linked to metabolism because they need energy to function. Specifically, they use one of the cell's basic energy molecules, nicotine adenine dinucleotide (NAD). They also regulate important biological processes, such as apoptosis (planned cell death), adipocyte and muscle cell differentiation, energy expenditure, and telomere length regulation. Because calorie restriction and life span extension are associated, one might reasonably assume that the cell organelles and processes associated with metabolism are involved. For example, mitochondria provide energy for cells, but they also produce free radicals that can cause oxidative damage to some biological molecules in a cell. This damage may directly affect aging, and mitochondrial dysfunction has been implicated in a number of age-related diseases, including Alzheimer's disease.

Eric Verdin and colleagues have contributed to our understanding of several of the sirtuins (Merksamer et al. 2013). Sirtuins were initially thought of as histone deacetylases, but SIRT3, for example, is a mitochondrial protein, and there are no histones in the mitochondria. However, mice lacking SIRT3 have many hyperacetylated proteins. Verdin believes that the acetylated proteins are primarily metabolic enzymes and that SIRT3 is a global regulator of metabolism in the mitochondria. He also believes that SIRT3 is directly involved in the relationship between caloric restriction and life span.

Verdin has also explored the regulation of SIRT1. This protein is highly expressed only in the cells of the testes and in embryonic stem cells, although its RNA is widely expressed throughout the body. The messenger RNA (mRNA—a copy of the DNA gene that is used to carry the information to where it is used) may be further regulated after the RNA is made but before the protein is made. The SIRT1 mRNA is targeted by several very small regulatory RNAs, called microRNAs. microRNAs are commonly used in this manner to regulate the amounts and activities of other proteins.

Some research has shown that resveratrol and other polyphenols activate SIRT1 and extend the life span of yeast (resveratrol competitively inhibits cAMP-degrading phosphodiesterases to increase intracellular levels of cAMP [Park et al. 2012]); resveratrol has also been shown to produce similar results with SIRT2. The upshot is that resveratrol can mimic calorie restriction in mice genetically engineered to become obese on a high-fat diet: the obese mice live longer when treated with resveratrol. Synthetic compounds that replicate the effects of resveratrol are in Phase I human trials. Unfortunately, other studies have not duplicated the positive results with resveratrol, and so, the value of that compound is not clear.

But there might be other drugs that do work to activate the SIRTs and provide the benefits of calorie restriction. For example, nicotinamide is a noncompetitive inhibitor of Sir2 and SIRT1. Increased expression of PNC1 (pyrazinamidase/nicotinamidase 1), which encodes an enzyme that deaminates nicotinamide, is required for life span extension by calorie restriction. Several other compounds are being investigated, and so far the results have been mixed. One interesting candidate is a synthetic compound, SRT1720, that potently activates SIRT1. SRT1720 increases oxidative metabolism in skeletal muscle, liver, and brown adipose tissue and so enhances endurance and strongly protects against diet-induced obesity and insulin resistance. Other candidates for investigation include rapamycin, green tea flavonoids, a histone deacetylase inhibitor, and microencapsulated curcumin.

Roundworms and Longevity

One of the favorite organisms for studying aging is the roundworm *Caenorhabditis elegans*. *C. elegans* normally matures through four larval stages before reaching the adult stage. However, if conditions aren't appropriate (e.g., too little food, temperature not right, overcrowding), these fascinating animals can slow their development. After the second larval stage, they can take a different course of development and transform into an alternative third larval stage called a dauer larva. Dauer larvae can live in that form for a considerable time until conditions improve.

Cynthia Kenyon has studied these worms. Her laboratory found that mutations in a particular gene called *daf-2* doubles the life span of worms (Kenyon 2011). This gene regulates the formation of dauer larva. *Daf-2* encodes a protein called the IGF-1 receptor. The receptor occurs throughout the animal kingdom. Furthermore, IGF-1 signaling is essential for regulating life span in many organisms. For example, mice lose this receptor as they age.

Another group of proteins is related to the insulin/insulin-like growth factor pathway. The Forkhead box O (FOXO) proteins are transcription factors that turn other genes on and off. They include *daf-16* in worms and dFOXO in flies.

Mice have four FOXO genes, but none have yet been shown to be involved in aging. Kenyon has studied *daf-16* (Libina, Berman, and Kenyon 2003). Two more genes that are downstream (or activated later in this particular genetic pathway) of *daf-1* and *daf-16* help the worms live longer. These genes might be part of a system that regulates dauer larva formation and life span. *Daf-16* also interacts with a kinase, an enzyme that adds a phosphate group to a protein. This modification is commonly used in biology to change the activity or structure of proteins.

Mitochondria and Aging

Mitochondria are the energy factories of the cell. They are where the food we eat is ultimately turned into the chemical energy that our cells need to sustain life. However, more recently, scientists have come to recognize the intimate role that mitochondria have in many key cell processes, some of which might be involved in aging.

For example, mitochondria have their own DNA, which is distinct from DNA in the nucleus. This mitochondrial DNA (mtDNA) is a remnant of the time when mitochondria were free-living bacteria-like organisms, before they began a sort of symbiotic relationship inside our cells. The mtDNA of older human individuals has more mutations than are seen in younger humans. Those mutations can cause deficiencies in cell respiration that result in signs associated with aging in mice. Those mutations might result from free radical oxidation by the byproducts of metabolism, but this remains to be proved. Mistakes by the mitochondrial DNA polymerase might also be involved.

In other studies, mutations in two genes, *isp-1* and *nuo-6*, greatly increased the life span of the worm *C. elegans*. These genes are involved in the energy reactions in mitochondria. If free radicals really do cause aging, the worms with these beneficial mutations should have lower levels of free radicals. Unfortunately, it does not seem to be that simple. Even more surprising, antioxidants actually reduced life span in the mutant worms. What could explain these unexpected results? One possibility is that the oxidizing compounds might signal other pathways to turn on, and those pathways might extend life span in the worms. This would be a very useful protective mechanism for aging organisms.

Worn-Out Parts

One set of theories about aging involves a buildup of molecular damage to the cells. The analogy is that the human body is like any other piece of machinery—constant use results in wear and tear that is never completely repaired. Over time, small injuries build up and interfere in more and more significant ways.

Humans clearly have the ability to repair some damage to tissues and organs. Wounds heal. Bones knit. We develop resistance to many infectious diseases. Still, even though we can repair some damage, aging brings on many changes that accumulate over time. The biochemical reactions in our cells and tissues are susceptible to mistakes and incomplete reactions. Proteins can become tangled or damaged in other ways, and the ability to clear those damaged proteins might be reduced over time. Tissues and organs accumulate damage: arteries become clogged by atherosclerosis, lung capacity is reduced, bones lose density, skin loses its elasticity, and brains, kidneys, eyes, ears, and other organs function less effectively. The immune response becomes less robust, compromising our ability to resist diseases. These changes might result from simple wear and tear.

Mitochondria and Reactive Oxygen Species

As noted earlier, mitochondria are the energy factories of the cell. In them, the food that we eat is converted by biochemical reactions into an energy-containing molecule called ATP. That series of reactions also produces some very reactive compounds that can bind to and damage proteins, lipids, and DNA. These are called reactive oxygen species (ROS). The free radicals that are often discussed in popular health literature are one type of ROS. ROS have been hypothesized to be involved in aging (Storz 2006). The antioxidants found in the food we eat can help counter the effects of ROS; even ketchup boasts of containing one type of antioxidant, called lycopene.

In addition to being produced as normal byproducts of metabolism, ROS can result from inflammation, infection, or consumption of alcohol and cigarettes. Damage can also come from exposure over time to various kinds of radiation, including X-rays, gamma rays, background radiation, and even sunlight. The DNA double helix is composed of two chains of sugars and phosphates connected by nucleotide bases. Radiation can break one or both of the backbone chains, or it can damage the nucleoside bases. Enzymes within the cell can repair most of this damage, but over time, those mistakes can accumulate to cause mutations.

Mitochondria undergo both fusion and fission. Dysfunction in this cycle has been suggested as a cause of neurodegenerative diseases, diabetes, cancer, and aging. Figge et al. (2012) used a computer-based modeling approach to study mitochondrial quality control. They found that the cycles of fission and fusion are critical to mitochondria and that these cycles are reduced in aging mitochondria.

D'Aquila et al. (2019) examined the association of epigenetics with mitochondrial aging by looking for patterns of methylation in a particular type of genes involved in mitochondrial maintenance. They found two sites that, when methylated, seemed to be associated with problems in the mitochondria. These are intriguing results, but clearly, these studies need to be extended to show that these sites are involved with aging.

Telomeres

Chromosomes are complex structures made up of the DNA that encodes genes and structural and regulatory proteins, such as histones and other proteins. Each individual human cell contains about 2 meters of DNA that is packed tightly into chromosomes. With that much DNA in each cell, maintaining control of the DNA during cell division and at other times is clearly important. Telomeres, the structures that form the ends of chromosomes, help with this task. Telomeres are made of DNA and have a specific sequence of nucleotides. The telomeres are somewhat analogous to aglets, the plastic tips that hold the ends of shoelaces together, though of course this is an oversimplification.

Longevity has been associated with telomeres (Chan and Blackburn 2004), but the mechanism isn't fully clear. One possibility involves cell division, a normal part of the cell's life cycle. Different types of cells divide more or less frequently; for example, the cells that form the skin, hair, blood, and the lining of the intestine are among those that divide the most rapidly. Every time a cell divides, the telomeres on the chromosomes in that cell get shorter (decreasing by somewhere between 30 and 200 base pairs). When they get too short, the cell ceases to be able to divide any longer, and it dies. For this reason, telomeres are associated with aging and cancer. Telomeres do not get shorter in the cells of tissues that do not normally divide.

Greider and Blackburn (1985) discovered an enzyme called telomerase that adds bases to the telomeres and keeps them from being worn down too quickly. However, as cells go through more divisions, they begin to use up their supply of telomerase.

Although telomeres have been described as "molecular clocks," there are no firm conclusions about whether shorter telomeres cause earlier death in people. For example, one study found that patients over sixty years of age with shorter telomeres were more likely to die from infection, disease, or heart attack, and some researchers have speculated that shorter telomeres may predispose to cancers (Bojesen, 2013). Others, however, have suggested that shorter telomeres might slow cell replication and so protect against cancer. The lack of agreement points to a common problem in biological research: researchers must sort through complex data to find evidence linking complicated phenomena.

Protein Homeostasis

Protein homeostasis is essential to a healthy cell—that is, the cell must be able to maintain its normal complement of proteins, which are continually made, used, worn out, and destroyed. Many things can damage proteins (e.g., ROS can cause conformation changes), and multiple cellular systems are involved in repairing damaged proteins and, for those proteins that cannot be repaired, degrading and removing them. The loss of protein homeostasis has been implicated in neurodegenerative diseases and aging.

Two primary systems are used to maintain protein homeostasis (Proctor and Lorimer 2011). First is the system that corrects protein misfolding. Proteins are tightly folded molecules, and under stressful conditions sometimes become misfolded. Molecular chaperones help misfolded proteins to resume their normal functional conformation. Specifically, misfolded proteins bind to the chaperone hsp90 and cause it to release the heat shock factor HSF-1. The free HSF-1 then forms trimers, migrates into the nucleus, and upregulates the production of other chaperones. Second, some proteins are damaged beyond the ability of the systems to repair them. Proteolytic mechanisms degrade those proteins so their components can be recycled or disposed of. There are two main protein degradation systems. In the proteasome/ubiquitin system, proteins to be disposed of are labeled with the protein ubiquitin, which causes them to be taken up by the proteasome, a cellular organelle that acts like a garbage disposal. Damaged proteins and other cellular materials are drawn into the proteasome, and the proteasome degrades them. The other system is autophagy, in which damaged proteins and organelles are transported to a compartment in the cell called the lysosome, where they are degraded. Both the proteasome/ubiquitin system and autophagy become less effective with aging (McAuley et al. 2017).

When proteins become damaged or misfolded, they have a tendency to aggregate, or clump together. Protein aggregation can overwhelm the protein homeostasis mechanisms and lead to cell death. Even during regular or "healthy" aging, protein aggregates tend to accumulate in older organisms. But neurodegenerative diseases are associated with aggregations of specific proteins. In Huntington's disease, the protein is huntingtin. In Alzheimer's disease, they are tau and amyloid beta. In Parkinson's disease, it is alpha-synuclein. When autophagy is inhibited, worn-out proteins can accumulate and interfere with the normal functions of important proteins, killing neurons and other cells (Finkbeiner 2020). Inhibition of autophagy has been implicated in neurodegenerative diseases and aging.

Cellular Senescence

For an organism to survive, cell growth must be carefully controlled. Cell proliferation is key to growth and development, but at some point organs and tissues are fully developed, and so proliferation slows or stops. When cell growth is uncontrolled, the result can be cancer. Cellular senescence is an important mechanism for preventing uncontrolled cell growth (Rodier and Campisi 2011). Senescent cells are cells that have lost their ability to divide but continue to be metabolically active. Senescence can be caused by numerous stressors, including genotoxic agents, irradiation, nutrient deprivation, hypoxia, mitochondrial dysfunction, and oncogene activation. Senescent cells also secrete increased amounts of substances, such as cytokines, growth factors, and proteases, that are involved in inflammation, wound healing, and growth responses; this collection of substances, which numbers in the hundreds, is known as the senescence-associated secretory phenotype. Over time, the number of senescent cells (called the senescence burden) increases (Neves et al. 2015).

Cellular senescence is a classic example of antagonistic pleiotropy. Early in life, it is beneficial and helps to maintain tissue homeostasis. It is essential for embryogenesis, in that it is involved in the development of tissues into specific forms and patterns as the embryo grows. Later in life, however, it is detrimental and associated with multiple age-associated diseases, such as cancer, Alzheimer's disease, Parkinson's disease, and atherosclerosis.

Cell Signaling

Communications between cells are critical for the health of cells, but those signaling pathways can become disrupted due to age. For example, the mTOR pathway (mentioned in Chapter 4) senses levels of amino acids, other nutrients, and signaling molecules such as growth factors, hormones, and insulin (Hansen et al. 2007; Albert and Hall 2015). The FOXO family of transcription factors is present in forms of life at all levels, from worms to humans, and they regulate life span (Kenyon 2010). Insulin and insulin-like growth factor are involved in energy regulation and growth (van Heemst et al. 2005). Transforming growth factor beta signals through other molecules to phosphorylate SMAD2/3 or SMAD 1/5/ 8 to control different pathways (Walton, Johnson, and Harriman 2017). Nuclear factor kappa B is a master regulatory molecule that controls multiple biological activities, including immunity, inflammation, cell differentiation, and apoptosis; it is associated with aging, and its inhibition slows aging (Tilstra et al. 2011). As noted in Chapter 4, aging is associated with increased levels of chronic inflammation and of circulating cytokines (Michaud et al. 2013). Interleukin-1, interleukin-6, and tumor necrosis factor alpha are associated with a greater risk of mortality, and the latter two are are associated with frailty. Inflammation may also be involved in various neurodegenerative diseases.

Future Directions

While molecular geneticists have not yet realized Ponce de León's dream of a Fountain of Youth, a great deal of progress has been made in understanding the process of aging. The knowledge gained in these studies may produce treatments that can relieve considerable suffering. However, none of these studies have examined the wisdom of extending human life. Is there an advantage to aging and death? Is there some degree of evolutionary pressure to end the life of organisms at some point? Some suggest that improving health span—the time we spend in good health—would have a more beneficial impact on society than just increasing life span.

In this volume, we will not dwell on the questions of philosophy that underlie these matters, such as what the world would be like if humans conquered death. But now that the Pandora's box of aging research has been opened, can it be closed again or controlled? Humans have some experience with knowledge and technologies that affect individual lives and societies as a whole, such as gene editing technology, and the results to date have been mixed. It remains to be seen what we do with this knowledge about aging and death.

Since the cells of humans and other animals innately have the ability to repair themselves, one might wonder if there might be a way to enhance these repair mechanisms to extend life. Today the primary foci of aging research are potential new drugs and regenerative medicine through stem cells.

New Drugs

Several studies have focused on mTOR, a substance that, as we have seen, regulates many key cellular activities, including cell growth, proliferation, motility, and survival, and protein synthesis (Weichhart 2018). It has been linked to aging in a number of experimental systems, though the mechanism is not fully understood. Substances that inhibit mTOR, such as rapamycin, may slow the senescence of human and mouse cells. In addition, rapamycin has extended the life of middle-aged mice in multiple studies. However, it may also have some toxicity (Barlow, Nicholson, and Herbert 2013).

As discussed earlier in this chapter, resveratrol may have some efficacy in slowing aging, though it seems less promising than initially thought; its mechanism of action is assumed to be mimicking calorie restriction. Kenyon has also reported a number of small molecules that forestall aging (Zhang et al. 2020).

In nematodes and mice, clioquinol inhibits the mitochondrial enzyme CLK-1, whose activity has been associated with aging. Finally, the antioxidant SkQ1 inhibits the involution of the thymus in rats (Obukhova et al., 2009).

Undoubtedly, many other drugs and drug targets will be identified in the future.

Repair by Stem Cells

As noted previously, the human body has considerable ability to repair damage to its cells. However, there are limits to that ability. The central nervous system and the heart can only repair some damage, and significant tissue loss (e.g., strokes, heart attacks, trauma) can rarely be repaired. Furthermore, cells' ability to repair themselves deteriorates with aging.

We can replace lost cells with new cells that arise from populations of stem cells. Stem cells are a special type of cell that can remain in an undifferentiated state indefinitely and can be developed into any other type of cell in the body. Embryonic stem cells are, as the name implies, found in the embryo; they are what makes it possible for a fertilized egg to develop into a new human. They divide and proliferate and develop into all of the cell types that form our bodies. Even after the human body is fully formed, however, some of these cells remain in an undifferentiated state and are available to produce new cells later on. As we've seen, certain organs and tissues continually regenerate—mostly those that are exposed to harsh environments or significant wear and tear, such as skin, blood, and the cells that line the intestines. The cells that provide this ongoing replenishment are called tissue-specific or adult stem cells.

Other organs, such as the brain and heart, have much more limited abilities to regenerate lost cells. In fact, for some time, it was assumed that these organs had no ability to regenerate. Recently, stem cells have been found for these organs too, but those seem to be more involved in minor maintenance and do not appear to have the ability to repair the large-scale cell death that occurs in heart attacks and stroke.

For years, biomedical scientists have wanted to leverage stem cells to regenerate body parts. That promise was brought much closer to realization by the discovery of induced pluripotent stem (iPS) cells by Dr. Shinya Yamanaka (Takahashi and Yamanaka 2006; Takahashi et al. 2007). Even differentiated adult cells can now be directly reprogrammed to become another cell type (e.g., skin cells can be directly reprogrammed to be beating heart cells). Reprogramming has vast potential. However, this exciting technology is still very new, and there's a lot we don't know. Exactly how are iPS cells and stem cells related? Are iPS cells stable? Are the epigenetic markers on adult cells derived from iPS cells as those on normal adult cells (e.g., DNA methylation patterns)? These and other issues are the subjects of furious research in laboratories around the world.

Of course, stem cells are not panaceas either. Like other cells, stem cells themselves might age. Signaling pathways might be overactivated and degrade stem cells over time. One example comes from the stem cells that make new blood cells. These stem cells make fewer new cells in older mice than in younger mice. Perhaps the older cells suffer from a shortening of their telomeres (as discussed earlier in this chapter), or their DNA might be damaged over time, or they may gradually lose the ability to divide. Second, stem cells are a part of their environment. They affect their local environment and are, in turn, affected by it. So perhaps aging cells around stem cells may lose their ability to secrete factors important for stem-cell vitality.

One of the defining characteristics of stem cells might also contribute to their aging: cell division. Stem cells continue to divide over time to renew themselves. However, DNA replication, which is required for cell division, is error-prone. It is possible that errors might accumulate in stem cells over time and increase the risk of tumor formation. Or tumor suppressor mechanisms elsewhere in the body might be turned on that could affect stem-cell division. The tumor suppressor gene p16INK4a is a good example of this possible mechanism. It regulates cell-cycle arrest and senescence for many types of cells, and more of it is expressed as cell age. It also hinders DNA replication after cells (e.g., brain or bone marrow stem cells) are exposed to radiation or other cancer-causing events.

8

Death of Cells

The human body is estimated to contain about 37 trillion cells (Bianconi et al. 2013), an almost inconceivable number. Each day, a lot of those cells die. Some die because of their function. For example, skin cells are exposed to the environment. They dry out, are exposed to sunlight, and suffer the wear and tear of ordinary living. Sometimes cells die when we don't want them to. Neurodegenerative diseases, such as Alzheimer's disease and Parkinson's disease, feature the death of brain cells. Critical cells can be lost in a heart attack or stroke or diabetes. The opposite can happen too: cells may continue to live when they should not, as happens with the uncontrolled growth of cancer.

So why do cells die? There are about 200 different types in our body, and each has a different function and a different average life span. White blood cells live for about thirteen days, and red blood cells live for about four months. Skin cells live about thirty days, and liver cells die after about eighteen months. Brain and heart cells last a lifetime.

Cells die for many different reasons. Some are killed by trauma. Burns, frostbite, and gunshot wounds all physically destroy cells. For other cells, the process is much more controlled. Programmed cell death (PCD) is critical for our reproduction, development, and survival. Several types of PCD occur. Apoptosis and autophagy are highly organized genetic programs. Necrosis is typically caused by external factors, such as infection or trauma, and is less regulated. Nectoptosis is a form of necrosis that is more organized.

Life and Death of Cells

How can we know when a cell is really dead? This question is similar to the question of when a person is dead (see Chapter 2). Obviously, cells lack a heartbeat, they don't breathe, and they cannot be checked for brain activity. So what is the best way of determining if they are dead? Several suggestions have been proffered, but all have flaws. For example, one possible way of determining cell death is to look for activation of the caspases, the enzymes that destroy cells in programmed cell death. However, caspases can be activated under other circumstances without causing cell death. Thus there is no clear biochemical test of cell death. Like in human death, cell death lacks a clear "point of no return."

The Biology of Death. Gary C. Howard, Oxford University Press. © Oxford University Press 2021.
DOI: 10.1093/oso/9780190687724.003.0008

In the laboratory, cells are identified as dead or alive with the use of dyes (e.g., trypan blue) that cannot pass through the intact plasma membrane of living cells. The membrane is compromised in dead cells, and so the dye enters the cell and can be easily seen with a microscope as a blue color. Living and dead cells can also be differentiated with propidium iodide or other dyes, which penetrate the compromised membrane of the dead cells. Once inside the cell, this type of dye binds to DNA and will fluoresce when exposed to light of the appropriate wavelength. The cells can then be sorted with a fluorescence microscope or by flow cytometry.

An international consensus suggested a morphological definition of cell death (Kroemer et al. 2009) based on any one of these criteria: loss of the plasma membrane, complete fragmentation of the cell and its nucleus, and/or engulfment of the cell and its remnants by another cell. A more recent consensus modified this definition somewhat: "irreversible degeneration of vital cellular functions (notably ATP production and preservation of redox homeostasis) culminating in the loss of cellular integrity (permanent plasma membrane permeabilization or cellular fragmentation)" (Galluzzi, Vitale, and Aaronson 2018).

How long does it take a cell to die? This is a more difficult question than it might at first seem. The simple diffusion of protein or other molecules across a cell takes some time. If diffusion were the only mechanism for the spread of death, then many biological processes would be much slower than they actually are. There must be another mechanism. There is, and it involves a phenomenon called trigger waves. Trigger waves allow the transmission of information rapidly over large distances (Gelens, Anderson, and Ferrell 2014). They involve positive feedback of the signal, and they do not slow down or lose strength over distance. Some examples are the propagation of an electrical signal along a nerve and waves of calcium or mitosis that spread across eggs.

Scientists at Stanford University (Cheng and Ferrell 2018) determined the speed of cell death in a type of programmed cell death, called apoptosis, in the eggs of the African clawed frog (*Xenopus laevis*). The frog eggs are very large cells (about 1.2 mm), and they are easily seen and manipulated in the laboratory. Eggs that are not fertilized die by apoptosis. The apoptosis pathway has several positive feedback loops. The researchers noted that the cells die far too rapidly for the normal methods of transmitting information, such as diffusion or flow. The trigger waves in the cells reached 30 mm per minute. So a very long cell, such as a neuron with an axon of 100 mm, might take over three minutes to die.

Cells in culture can live for very long times, especially if manipulated to proliferate indefinitely (immortalized cells). The most famous immortalized cells are those of the HeLa line (Lucey, Nelson-Rees, and Hutchins 2009). The original cells from which the line was established were collected from an African American terminal cancer patient named Henrietta Lacks by Dr. George Gey at the Johns Hopkins Hospital in 1951. These were the first cells to be grown successfully in culture and for seventy years have been used by scientists around the world for a whole range of studies. However, the cells were removed without the patient's knowledge or consent, a common practice in that era. Ms. Lacks died from cervical cancer, and her family received no benefit from the contributions her cells have made to science. Her story was told in great detail in the 2010 book *The Immortal Life of Henrietta Lacks* by Rebecca Skloot.

As we have seen in previous chapters, most cells do not divide after the tissue or organ has reached its normal full growth. At that point, growth-suppressing proteins and microRNAs, which are short RNAs that hinder protein synthesis and cell division (Sheetz 2019). There are times when cell division needs to turn on, however. Wound healing is one. At other times cells lose control and begin dividing inappropriately, which is how we define cancer. The cancer cells have become transformed and will divide indefinitely, even if placed in culture. This is quite different from normal cells, which will divide a relatively small number of times and then die off.

Cellular Aging and Death

Different groups describe cell aging and death in different ways. For example, Alberti and Hyman (2016) see three aspects. First are the causes of cellular aging (including the loss of proteostasis (maintenance of normal protein levels) and the reduction of telomere length). Second are ways that cells compensate for or try to prevent aging. Third, the integrative hallmarks include changes in cell communication and the exhaustion of stem cell populations. López-Otín et al. (2013) looked at cellular aging and death in a different way. They cited nine causes, including genomic instability, loss of telomeres, changes to epigenetics, cellular senescence, stem cell exhaustion, loss of protein homeostasis, inability to sense nutrients, mitochondrial dysfunction and changes to intercellular communications. These factors might also work together in some way to cause cells to age and die.

Cell death might be caused by the accumulation of cellular damage. In young cells, the damage can be repaired, but older cells begin to lose that ability. For example, damage to the cell's biomolecules, such as proteins and nucleic acids, may lead to a diminished homeostatic capability. Alternatively, metabolic capability may be diminished. A question always looms over such discussions: are these changes are a *cause* of aging or a *result* of aging?

Another possible mechanism for cellular aging is dysfunctions in the cell's proteins (Hipp, Kasturi, and Hartl 2019). Cells manage the health of their proteins by maintaining a network of molecular chaperones and the proteolytic machinery that can remove and recycle damaged proteins. This process is called proteostasis. Many age-related diseases are associated with dysfunctions in proteostasis. For example, most of the major neurodegenerative diseases feature dysfunction of proteostasis and aggregations of a specific protein.

As discussed in Chapter 7, telomeres—the repeated short sequences of DNA at the ends of chromosomes—may be another mechanism involved in cell death (Blackburn and Gall 1978). The telomeres seem to function to stabilize the chromosome ends. Each time the cell divides, the telomeres lose a number of the repeated DNA sequences. An enzyme called telomerase can repair and replenish the telomeres, but only up to a point: in mice and humans, the number of repeats declines with age, and the telomeres shorten.

Another possibility might be dysfunctional phase separations (Alberti and Hyman 2016). The cytoplasm is organized into membrane-bound organelles and also membrane-less compartments that involve phase separation. Each allows spatiotemporal control of reactions, but phase separations are sensitive to changes in concentration, pH, and energy levels. When these separations go awry, cell death can result.

The death of specific types of cells can stress other cells. For example, diabetes is caused by the loss of beta cells in the pancreas that secrete insulin. Diabetes, in turn, renders the cells in other organs susceptible to stresses from too much glucose in the bloodstream.

Injury and Disease

When things go wrong, cells in organs and tissues die, but the entire organism does not necessarily perish. Some cells die from injury or disease. Others die because of PCD. The function of those cells is lost, sometimes irreversibly, and the results are often devastating. For example, cell death in the brain or heart is serious. Unlike liver cells, which can easily regenerate, the cells in those organs are postmitotic—they have exited the cell cycle and no longer divide to create new cells. For a long time, no stem cells were known for the brain and heart, but in more recent years, a limited number of stem cells were found.

The death of some cells is normal in aging. For example, hearing loss is a common feature of aging. In many cases, it is caused by the loss of the sensorineural cells in the ear (called hair cells). Those cells die by both necrosis and apoptosis (Morrill and He 2017). The loss of hearing is not without consequences, but injuries and diseases cause the death of cells that can result in more serious consequences. In some cases, the patient survives and recovers partial or complete function. In other cases, the loss of function is permanent and can be extremely debilitating. Several other mechanisms exist, including autophagic cell death, parthanatos, and more (Galluzzi, Vitale, and Aaronson 2018).

Loss of Oxygen

Cells depend on a continuous supply of oxygen to continue the biochemical reactions that provide the energy for life to continue. When that supply is lost, the tissues and cells begin to die. For example, during a stroke, the blood supply to a portion of the brain is cut off, either when an artery ruptures or an artery is occluded by a clot. Neurons need a lot of energy, and they depend on a continuous supply of oxygen and glucose to make that energy. Without it, they begin to die within minutes. PCD is deeply involved in the processes initiated by a stroke.

A similar problem occurs if the heart is deprived of blood. Blood flow and thus the oxygen supply to parts or all of the heart can be disrupted because of a blood clot or an arrhythmia. The muscle cells rapidly run out of oxygen and begin to die. The resulting myocardial infarction (heart attack) will be fatal unless the blood flow is restored within minutes.

In gangrene, blood flow is cut off. Without a supply of oxygen, the tissue dies. There are two types of gangrene. In dry gangrene, the tissue dies, turns green-black, shrinks, and dries up. In wet gangrene, bacteria infect the dying and dead cells, and the tissue swells and smells. Gangrene may result from an infection, an injury (e.g., wound, burn, frostbite), or a chronic disease (e.g., diabetes, peripheral artery disease). Fournier's gangrene a type of dry gangrene, (Thwaini et al. 2006) is caused by bacterial infections (e.g., coliforms, *Klebsiella*, *Streptococci*, *Staphylococci*, *Clostridia*, bacteroids, and corynebacteria). It occurs mostly in diabetics and those who chronically abuse alcohol. The infection releases exotoxins and enzymes (e.g., collagenase, heparinase, hyaluronidase, streptokinase, and streptodornase) that cause microthrombosis of the small subcutaneous blood vessels, and the tissue damage and loss of oxygenation .

Cell death seems to follow the loss of oxygen in two waves. The immediate death seems to be due to necrosis. Days later, cells in the area of the stroke die from apoptosis. Levels of the autophagy marker LC3 also increase during a stroke, indicating that autophagy may also have a role. Interestingly, many of the genes involved in these processes (so-called death pathways) are on the X chromosome and thus may show sex differences (Chauhan, Moser, and McCullough 2017).

Apoptosis is seen in newborns after birth asphyxia, sudden death, and white matter injury. Apoptosis is also seen after other injuries, such as spinal cord injury, and is sometimes seen at sites distant from the initial injury. Premature infants experience many challenges as they struggle to transition from the protected environment of the uterus. Many of their organs are not sufficiently mature to function properly. They typically have problems with their lungs, heart, and digestive tract. Their lungs, in particular, are often not developed sufficiently to absorb enough oxygen. The inability of the lungs to fully oxygenate the blood may further weaken the fragile intestinal lining and allow those cells to be attacked by bacteria. Necrotizing enterocolitis (NEC) is a very serious infection that occurs in about 1 in 2,000–4,000 premature infants (Neu and Walker 2011). NEC was sometimes seen in the 1960s and before, but it did not become common until advances in modern neonatal care; the improved survival of ever-younger babies resulted in more babies with severely underdeveloped intestines. NEC has a prevalence of about 7 percent in babies who weigh between 500 and 1,500 grams, and the likelihood of death is 20–30 percent. This infection can damage or destroy the wall of the intestine. In the worst cases the intestine can perforate, allowing the seepage of stool into the baby's abdomen; if that happens, the infection can become overwhelming. The accompanying inflammation can attack the brain. For those who survive, scarring and narrowing of the intestines can cause later problems. If part of the intestine has to be removed, the infant may be unable to absorb nutrients properly.

The actual cause of NEC is unknown, but several predisposing factors contribute to the disease. There may be a genetic component to the susceptibility. Polymorphisms are found in the toll-like receptors that recognize microorganisms and activate the innate immune system (Niño, Sodhi, and Hackham 2016). The intestinal barrier function is inadequate. Less mucus is present, the junctions between the cells that line the intestine are not complete, and so the permeability is high. Differences in microbiota may contribute to susceptibility; the diversity of microbiota in affected neonates is lower than in babies not affected, but no specific species of microorganism has been identified as especially problematic. Gastric acid secretion is low. Levels of IgA are lower than in full-term infants. The inflammatory response is quite high in the cells that line the infant intestines, and the levels of chemokines and cytokines are also elevated.

Autoimmune Diseases

The immune system is a vital defense against multiple pathogens, but it must be carefully regulated. If that regulation is disrupted, cells can be injured or killed. For example, diabetes involves the inability of the body to regulate blood sugar levels. Sugars from our diet are absorbed from the blood with the use of a hormone called insulin. Insulin is made in and secreted by the beta cells of the islets of Langerhans in the pancreas. Diabetes results when the beta cells are lost or can no longer produce insulin or when other cells in the body lose their ability to use insulin (insulin insensitivity).

There are two major types of diabetes. Type 1 (or juvenile-onset) diabetes usually manifests early in life. In type 1, the pancreatic cells are killed by an autoimmune reaction. Studies have been limited by the fact that beta cells are rare and buried deep in the pancreas. What triggers the autoimmune response is not known. What we do know is that lymphocytes invade the islets of Langerhans before clinical symptoms are noted (Campbell-Thompson et al. 2016). The actual destruction of the beta cells is caused when the immune system's T cells initiate apoptosis. Apoptosis is intimately involved with many aspects of diabetes (Lee and Pervaiz 2007). In mice models of diabetes, mice that lack certain types of T cells do not develop type 1 diabetes. Pancreatic cells of newly diagnosed diabetes patients have more expression of the apoptosis-associate protein Fas (CD95) on their surfaces than cells from people who do not have diabetes.

Type 2 (or adult-onset) diabetes typically appears later in life and is much more common than type 1. For some time, it was assumed to be a metabolic disease, but more recent research has shown that it has an autoimmune component that involves apoptosis (Thomas et al., 2009).

Neurodegenerative Diseases

Neurodegenerative diseases include Alzheimer's disease (AD), Parkinson's disease (PD), Huntington's disease (HD), amyotrophic lateral sclerosis (ALS), and more (Vila and Przedborski 2003). All are invariably fatal, and there are no drugs that change the course of the disease for any of them.

Although these diseases are distinctive illnesses, they share some features. All involve the destruction of specific types of neurons, aggregations of proteins, and ultimately death. Neurons are lost in each of these diseases, but the types of neurons that are lost differ. For example, AD results from loss of medium and large pyramidal neurons in the hippocampal regions. HD results from the loss of medium spiny neurons in the striatum. PD results primarily from the death of dopaminergic neurons in the substantia nigra (Moore et al. 2005).

Protein aggregations are found in many of the neurodegenerative diseases. Each disease features aggregations of specific proteins (with AD, it's beta-amyloid and tau; with PD, it's a-synuclein; with HD, it's huntingtin). These aggregations have long been suspected in disease, with the assumption being that the aggregates interfere with the synapse or other critical functions. Since then, a great deal of effort has been made to eliminate the aggregated proteins, but no benefit to patients has been seen. In fact, some research has pointed in the opposite direction and implicated single copies of the proteins called monomers or very small complexes of two or three monomers as the real culprits in disease.

That theory was strengthened by the finding that the aggregates in HD were actually protective—cells that developed the aggregates survived longer than those that did not. An analogy might be to asbestos, which years ago was commonly used as insulation in buildings but which we since have learned causes cancer when the small asbestos fibers become airborne and are inhaled. Asbestos that is on the pipe or ceiling but intact does not pose a problem; the danger comes when the asbestos is disturbed and friable pieces are released into the air. In like manner, the cell puts the mutant protein into a clump to store it safely away from the rest of the cell. If this theory is correct, then ironically, treatments that seek to disrupt the aggregates and eliminate the protein from the cell may be counterproductive. Once the aggregate is broken up, small, possibly toxic fragments may be released, harming the cell.

Programmed Cell Death

The human body contains an enormous number of cells, but sometimes certain cells have to be removed. In some cases, such as development or metamorphosis, new structures need to be formed. In other cases, cells are damaged or diseased and must be eliminated for the overall health of the host. Getting cells to their correct position during development and maintaining the organization of tissues and organs while they are functioning require carefully regulated control systems. Thus organisms have developed biological processes to kill off specific cells. In this way, life has commandeered death for its own purposes.

The first observations of this more controlled process were made in species that undergo metamorphosis from a juvenile form to an adult form. Developmental biologists have known since the 1800s that lots of cells are lost during development. However, the reason for the loss was unexplained until Glücksmann (1951) suggested that this cell death was part of the developmental program. In the mid-1960s, Richard Lockshin and Carroll Williams at Harvard observed that the larval abdominal muscle cells of American silkmoths died during the development of the larva into an adult (Lockshin 2016). Those muscles help the larva to move, but once the larva has changed into an adult, they have one last function: to push hemolymph (the moth's blood) into the veins of the wings to cause them to expand to their full size. Once this is accomplished, the muscle cells die. Lockshin and Williams coined the term "programmed cell death" to describe this phenomenon. They also showed that this was a cell-autonomous process, meaning that the cell itself caused its own death. Cells use common groups of genes or pathways to control PCD, and those genes are found in virtually every form of animal life, from worms to humans (Baehrecke 2003).

For some time, it was assumed that PCD appeared early on in the development of life on Earth, with the evolution from unicellular organisms to multicellular ones. A multicellular organism requires considerable organization to position all of its cells, tissues, and organs so that they work together and function appropriately. When some cells lose the ability to do that, the organism needs a mechanism to fix the problem, and the ability to eliminate those cells would be beneficial. Thus, it is easy to imagine how such as system would evolve in multicellular organisms. However, a bigger surprise is that a similar system also occurs in unicellular organisms. We will return to this issue in Chapter 11.

PCD is critical in several ways. For an organism to develop from an embryo to an adult, specific cells must live and others must die. The immune system must be refined so that it doesn't release antibodies that would attack our own tissues; this is done by selectively killing off particular cells. And of course PCD is an important component of the defense against illness—under certain disease conditions, cells "commit suicide" by turning on pathways that lead to death. PCD provides a mechanism by which these deaths can occur in an organized and controlled manner.

PCD includes apoptosis, necrosis, and autophagy. These are the main three, but there are more. Apoptosis prepares the cell's contents for recycling and use by other cells. Necrosis involves the death of cells by toxins or trauma and is generally less organized. It also prompts an immune and inflammatory response that can sometimes be problematic. Some researchers are reluctant to include necrosis in PCD because it seems less controlled. Autophagy is another form of PCD that allows the organism to optimize available resources by recycling damaged and worn-out proteins, organelles, and other cellular constituents. The cell activates this process when resources are limited.

Much of what is known about PCD was discovered in the nematode *C. elegans* (Conradt, Wu, and Xue 2016), which has several features that make it an ideal model system. *C. elegans* consists of only about 1,000 cells. Its genome is only 100 million base pairs, but surprisingly, it has about 20,000 genes, roughly the same number as humans. In fact, many of those genes are similar in both worm and human. The worm's developmental stages are well defined. And PCD can be directly observed in the worms by differential interference contrast (Nomarski) microscopy.

The regulation of PCD, even in *C. elegans*, is complex. A series of networks lead to expression of the genes *egl-1* and *ced-3*. One of the substances that is produced, CED-3, is a caspase protease that when activated sets in motion multiple events, such as fragmentation of the nuclear DNA, elimination of mitochondria, relocalization of the signaling molecule phosphatidylserine to the cell surface, and clearance of dead cells. Apoptosis

Apoptosis was first observed by the German scientist Karl Vogt. In 1842, he was studying the development of the midwife toad (*Alytes obstetricans*) in the laboratory of Louis Agassiz and noted that cells in the notochord (a cartillagenous embryonic structure that supports the organism) disappeared as the vertebrae appeared. In 1885, another German scientist, Walther Flemming, reported the loss of nuclear material in dying cells. We now recognize this loss as a critical element of apoptosis. The term *apoptosis* was coined in 1972 (Kerr, Wyllie, and Currie 1972). Amazingly, every cell is programmed to commit suicide. Under normal circumstances the mechanisms are held in check, but are ready to go if needed.

Examples are easy to cite. In human development, the hands and feet begin as a sort of paddle-like mass of tissue. Cells in the webbing between the fingers and toes are caused to die by apoptosis, allowing the fingers and toes to emerge as separate. When apoptosis is incomplete, the hand or foot has webs of tissue between the fingers or toes, a condition called syndactyly. Another example is that the brain initially makes many more cells than are needed. The cells that do not make appropriate connections with other cells are eliminated by apoptosis. A third example is the immune system. It is essential that the immune system learn how to differentiate the body's own molecules from those of invading entities so that it does not attack its own tissues and organs. This is accomplished in the thymus and other locations. Immune cells each secrete only a single type of antibody, and those cells that secrete antibodies against self are killed by apoptosis. When this process is incomplete, any leftover antibodies against self will result in an autoimmune disease. Apoptosis is also important for controlling the immune response. A rapid and overwhelming response to a pathogen is needed to control the pathogen. However, once the pathogen is destroyed, the immune system must be throttled back, and the unneeded immune cells must be eliminated.

Apoptosis is used in several ways to protect the organism. Some precancerous cells might not be caught by the immune system. In those cases, immune cells can provide the signals externally and induce apoptosis to kill the damaged cell. In other cases, cells can be infected by viruses. If those cells were allowed to survive, the virus could use the cell's machinery to replicate and infect more cells. The cells with the virus inside them must be killed in order for the rest of the body to survive. In like manner, cells with damaged DNA also need to be eliminated.

An average human loses 200 billion to 300 billion cells every day, and most of those are lost by apoptosis (Arandjelovic and Ravichandran 2016). It's essential for the cell's homeostasis that the debris from these apoptotic cells be cleaned up. One of the major cell types engaged in this cleanup effort is the macrophage. Macrophages engulf cell debris and degrade it. Macrophages are produced from immune stem cells in each type of tissue, which differentiate into tissue-specific macrophages, such as peritoneal macrophages in the peritoneal cavity, Kupffer cells in the liver, alveolar macrophages in the lung, and microglia in the brain. These cell types are sometimes referred to as "professional cleaners."

The macrophages do not just attack any cell they come upon. The apoptotic cell "identifies" itself to the macrophage by releasing signal molecules. These molecules may include nucleotides (e.g., ATP, UTP), the chemokine fractalkine, and the lipids lysophosphatidylcholine and sphingosine-1-phosphate. Receptors on the surface of the macrophages recognize the signals. In a parallel manner, healthy cells show specific molecules on their surfaces, such as CD47 and CD31, that direct the macrophages to leave them alone.

Other types of cells, such as fibroblasts and epithelial cells, also participate in the cleanup of apoptotic cell debris. While they are not as efficient as the professional cleaners, they are important in tissues lacking large numbers of macrophages, such as the airway and intestinal epithelia. Some more specialized cells are also involved. For example, the Sertoli cells within the testis remove apoptotic sperm cells. And retinal pigment epithelial cells remove the shed outer segments of photoreceptors in the eye.

Apoptosis takes two forms. Intrinsic apoptosis is initiated within the cell itself. Extrinsic apoptosis is initiated from outside the cell via extracellular ligands, appropriately called death receptors. The two types are essentially independent of each other, except that they converge on the same enzyme cascades. In either case, apoptosis involves a fairly consistent series of steps.

First, cells shrink. At the molecular level, apoptosis involves the major contractile proteins actin and myosin (Ndozangue-Touriguine, Hamelin, and Breard 2008). Actin forms a ring around the membrane and, with myosin, contracts to form the membrane protrusions (known as blebs) that characterize apoptosis. Other protrusions then form, such as microtubules. The DNA is cleaved into pieces. The cell breaks into small pieces (apoptotic bodies), each surrounded by a membrane containing the lipid phosphatidylserine. Macrophages sense the lipid and engulf those small pieces, and the contents are recycled for use by other healthy cells. cytoplasmic vacuolization and mitochondrial swelling

The primary enzymes involved in apoptosis are proteases called caspases (the name *caspase* is a short form of the full name: cysteine-dependent aspartate-directed proteases). At least fourteen caspases are known, and they can be classified by their general function as initiators, executioners, or inflammatory proteins. The words describe their functions. The initiators begin the process, and the executioners kill the cell. The inflammatory proteins characterize a related process called pyroptosis. However, caspases don't work alone. They function in a cascade, with one activating the next; more and more proteases are liberated with each step of the cascade. Apoptosis involves caspases 2, 8, 9, and 10 as initiators and caspases 3, 6, and 7 as executioners.

The caspase cascade and apoptosis are controlled by a large cast of characters (see Green 2005 for an excellent review). Most of these proteins are normally inside the mitochondrial membrane, but certain stimuli can cause them to relocate to the cytoplasm or nucleus, where they turn on caspases and nucleases. One of those proteins is cytochrome c, which activates caspases when released from the mitochondria. During intrinsic apoptosis, cytochrome c binds to other proteins, such as APAF-1, to form complexes that activate caspase-3. Caspase-3 and the other executioner caspases lead to the changes associated with apoptosis, such as chromatin condensation, DNA fragmentation, nuclear membrane breakdown, movement of phosphatidylserine, and formation of apoptotic bodies. The mitochondria also release other proteins, including 57-kD apoptosis-inducing factor (AIF). When apoptosis begins, AIF moves from the mitochondria to the nucleus to initiate chromatin condensation and DNA fragmentation. Endonuclease G, a nuclease encoded by nuclear DNA, is released from the mitochondria, travels to the nucleus, and begins nucleosomal DNA fragmentation. Cytochrome c and other apoptosis-related proteins are controlled by the Bcl-2 family of proteins: Bcl-2 and Bcl-XI inhibit apoptosis, while Bax and Bak induce it by making the mitochondrial membrane permeable so that it releases cytochrome c. BH3-only proteins are also involved in apoptosis, but the mechanism is not clear (Happo, Strasser, and Cory 2012); one possibility is that some BH3 proteins (e.g., Bad, Bik, Hrk) bind to anti-apoptosis proteins and release pro-apoptosis proteins, such as Bax and Bak, and another is that BH3-only proteins bind to Bcl-2 proteins, making them unable to prevent Bax or Bak proteins from being activated and setting apoptosis in motion.

Apoptosis proceeds smoothly hundreds of thousands of times a day in every human being, but occasionally something goes wrong. For example, in cancers such as lymphomas, the balance between the production of new cells and the loss of old cells is disrupted, resulting in too much growth. The overproduction of one of the apoptosis-related proteins, Bcl-2, may be part of the reason for this imbalance. And uncontrolled apoptosis can result in cases of ischemia and neurodegeneration (Czabotar et al. 2014).

Autophagy

Many cells have a normal life span, after which they simply wear out, die, and are discarded. Some cells, especially those exposed to the external world (e.g., skin cells), wear out with that exposure and must be replaced. Most organisms, including humans, have sophisticated systems to remove dead cells. Those systems are very important for maintaining health, and they are well used. Hundreds of billions of cells die every day in a human body, many because they have been damaged by normal physiological processes or interactions with pathogens. Autophagy is deeply involved in many of these processes (Shibutani et al. 2015).

Autophagy was first described in electron microscope studies in the 1950s and 1960s (Clark 1957; Ashford and Porter 1962; De Duve and Wattiaux 1966). There are three types of autophagy: chaperone-mediated autophagy, microautophagy, and macroautophagy. Chaperon-mediated autophagy has only been found in mammals. It uses specific receptors to selectively transport individual proteins from the cytoplasm across the lysosomal membrane into the lysosome. Once in the acid environment of the lysosome, they are degraded by the lysosomal enzymes. The major chaperone is heat shock protein (HSP) 70, which is deployed in times of stress, such as high temperatures. The chaperone binds to a specific protein and escorts it to the receptor on the surface of the lysosome.

In microautophagy, cellular components are captured by a portion of the lysosomal membrane that folds in to form a pouch, then pinches off and forms a vesicle inside the lysosome. The entire vesicle is quickly degraded by the lysosomal enzymes.

Macroautophagy (referred to as autophagy hereafter) is a well-understood process (Rubinsztein, Shpilka, and Elazar 2012). Autophagy-related genes were initially discovered in yeast, but analogous ones are found in humans. Autophagy is used by the cell in several critical functions. For example, it is important for defending the cell against pathogens, and it also helps to regulate immune responses by slowing the immune reaction after a pathogen has been dealt with.

Autophagy is highly regulated at the transcriptional, translational, and post-translational levels. The associated proteins function in several complexes—for example, the ULK complex recruits the P13 kinase complex, which is required for autophagy. The effectiveness of autophagy and the expression of autophagy genes, such as *ATG5* and *ATG7*, decrease with aging. The result is that damaged and senescent cellular components accumulate in the organism as it ages.

Although autophagy is regulated by multiple factors, such as the kinase JNK-1, the main factor controlling autophagy is mTOR (mammalian target of rapamycin). mTOR controls many processes involving metabolic homeostasis through a complex called mTOR complex 1 (mTORC1), which connects the internal workings of the cell with environmental signals regarding nutrients and growth factors. When conditions are favorable for growth (indicated by growth factor signaling through the insulin receptor), mTORC1 encourages growth and inhibits autophagy. When conditions are unfavorable and nutrients are lacking (e.g., reduced levels of oxygen, low cytosolic ATP levels), autophagy degrades and recycles the cell's own components. If pathogens have invaded or damaged a cell, autophagy can cause the cell to commit suicide in programmed cell death.

Autophagy involves a number of actions and proteins. With the help of ATG proteins, the phagophore begins to form from the endoplasmic reticulum

(ER). The ER is a network of membranes inside the cell, and the phagophore is a membranous extension of the ER. Several proteins, including ATG5, ATG12, ATG16L, and LC3 cause the nascent phagophore to enlarge. Next, the material to be degraded is located and captured. The expanding membrane surrounds the target to form an autophagosome that fuses to a lysosome and releases its contents into the lumen of the lysosome. The acidic environment of the lysosome and its degradative enzymes digest the cargo, and the resulting raw materials are then recycled for use.

Necrosis and Necroptosis

While apoptosis involves a careful program of well-defined events, necrosis is generally thought of as accidental, occurring as the result of trauma, infections, cancer, infarctions, inflammation, and other conditions. For example, necrosis can be caused by a loss of blood supply—say, from damaged blood vessels or poor circulation. In frostbite, water within a cell expands as it freezes, causing the cell to rupture and die by necrosis. Some spider and snake venoms contain enzymes that attack the cells and kill them. Necrosis is also seen in non-inflamed regions of the colon's epithelium, so it is not clear if necrosis is only associated with disease or might be a normal part of PCD in the intestines.

In apoptosis, cells shrink and divide into small packets. By contrast, in necrosis the cells swell and break open, releasing their contents. Damage to the plasma membrane allows water to enter the cell. The cell swells, and the cellular contents leak out. Degradative enzymes in lysosomes also escape, adding to the cellular damage. The cell contents include signaling molecules that elicit an immune reaction, leading to inflammation. Other signals attract phagocytes to clean up the cellular debris left by necrosis. The arrival of the phagocytes is a double-edged sword—while they can clean up debris, they also release molecules that damage surrounding cells and complicate the healing process.

A process called necroptosis seems to be a middle ground between apoptosis and necrosis (Berghe et al. 2010). Although necroptosis resembles necrosis more, it nevertheless features a more organized mechanism of cell death. Necroptosis and apoptosis share several components of their regulatory pathways; for example, TNFa signals in both pathways.

Necroptosis might occur when apoptosis fails for whatever reason. What starts it is not clear, but researchers have suggested, including TNF, ligation of Fas, and the involvement of innate immunity sensors. Like necrosis, necroptosis does not depend on caspases. However, unlike in necrosis, permeabilization of the cell membrane is controlled. Necroptosis begins when tumor necrosis factor (TNF)a binds to its receptor. This recruits the TNFa receptor-associated death domain and other factors that form the necrosome. The ominous-sounding death domain is a part of the receptor that attracts the caspases. Then another protein begins to breakdown the cell membrane, which allows the contents of the cell to spill out. That material attracts immune cells to clean up the debris.

Necroptosis is associated with disease conditions that involve tissue damage, such as heart attack and stroke, and diseases such as atherosclerosis, inflammatory bowel disease, neurodegeneration, and cancer. Because of this involvement, necroptosis is a target for therapies for these diseases.

Other Forms of PCD

A number of other forms of PCD occur, often in specific tissues or under specific conditions. *Pyroptosis* occurs in response to pathogens and heart attacks. It is different from other forms of PCD due to the different caspases (including caspases 1, 4, and 5) involved. They cause the cell to swell and break open. The released material, such as potassium ions, induces an immune response and inflammation. Multiple proteins sense the potassium ions from a cell and form a complex called an inflammasome that activates caspase-1 (not caspase-3, as in apoptosis). Caspase-1 is also known as interleukin-1-beta-converting enzyme (IL-1B), which causes fevers. Pyroptosis is involved in HIV infections (Doitsh et al. 2014).

Cornification occurs in the epidermis and is essentially the final step in the differentiation that forms the outer protective skin layer. The resulting corneocytes contain proteins, such as keratin, and loricrin, and various lipids; caspases are deployed to a limited extent. Proteins are cross-linked by transglutaminases, and lipids are released into the extracellular space to insulate the body from the environment. In essence, then, the outermost layer of the skin is composed of dead cells.

Mitotic catastrophe occurs during or just after failed mitosis—when a cell attempts to divide to form two identical copies of itself, but something goes wrong with the process, often because of damage to the cell's DNA. It features micronucleation and/or chromosome fragments. It can lead to apoptosis or necrosis.

Anoikis occurs in multicellular organisms where specific types of cells interact with the extracellular matrix to determine how much to grow and where to be (Gilmour 2005). For example, intestinal cells normally grow only in proximity to the structures of the intestine and other intestinal cells. When cells get out of position, such as when they lose their anchorage or when they attach inappropriately to a surface, apoptosis is initiated to remove the excess or mislocated cells. The signaling involved differs depending on the type of cell and the adhesion, but the integrin family of proteins is involved—humans have at least twenty-four integrins, and different cell types have different integrins. The integrins keep cells alive by regulating the signaling pathways that control apoptosis. Loss of adhesion sends signals that activate most of the BH3-only proteins and thus begin apoptosis.

Excitotoxicity occurs in neurons. When the neurons are exposed to excitatory amino acids (e.g., glutamate), the N-methyl-D-aspartate Ca^{2+}-permeable channel opens, cytosolic calcium ions go out of control, and lethal signaling pathways are activated. This pathway may be important in Alzheimer's disease. Excitotoxicity has a lot of overlap with apoptosis and necrosis and may actually be a variation on one or the other.

Wallerian degeneration is not yet very well understood, but we do know that it causes part of a neuron or axon to degenerate without affecting the cell body.

Paraptosis is triggered by IGF-1. It features cytoplasmic vacuolization and mitochondrial swelling but lacks the hallmarks of apoptosis. Paraptosis is initiated by members of the mitogen-activated protein kinase family (mitogens are small proteins that initiate mitosis or cell division) and is not inhibited by caspase inhibitors or anti-apoptotic Bcl-2-like proteins.

Pyronecrosis is similar to pyroptosis but differs in that pyroptosis requires the actions of caspase-1. Pyronecrosis does not. Instead, it involves the cathepsin-dependent cell-death pathway.

Entosis was first described in Huntington's disease. In it, cells begin to engulf their neighbors. Cells become susceptible to entosis when they lack adhesion to the extracellular matrix or during periods of glucose starvation.

Ferrroptosis depends on the presence of iron and results in the oxidation of lipids. It is involved in tumors, nervous system diseases, kidney injury and blood diseases. Its specific mechanisms are still unknown (Li et al., 2020).

Anastasis: A Near-Death Experience for Cells

Once cells initiate apoptosis, is the death of the cell inevitable, or can the process be stopped and reversed? For some time, the scientific consensus was that once apoptosis was triggered, the process inevitably went through the full sequence, and the cell died; there was no turning back. However, more recent research shows that might not be the case. Apoptosis can be reversed, even at late stages after caspase release and mitochondrial fragmentation. The recovery of cells after apoptosis has been initiated is called anastasis (Sun and Montell 2017; Tang and Tang 2018, Gong et al. 2019). It can occur after a range of cellular threats, including cold shock, toxins, and protein starvation.

When the process of apoptosis fails, fewer caspases are released, and not all the cell's mitochondria begin to leak cytochrome c at the same time, with the rest remaining intact. When this happens, apoptosis ends. The cells store away mRNAs before apoptosis begins, and those are used to restart normal cell processes. In anastasis, more than 1,000 genes are upregulated. But survival does not come without a cost to the cells. Cells that begin anastasis at late stages of apoptosis might not be completely healed. They may suffer from chromosome defects that can result in malignancy later. This may be especially relevant to cancer treatment, as treatments involving radiation and many chemotherapy drugs kill cancer cells by initiating apoptosis (Sun and Montell 2017). Some dying cancer cells may escape the treatment and begin anastasis. It is even possible that anastasis may promote metastasis.

While failed apoptosis is widely recognized, not everyone is convinced that anastasis is real. Obviously, it is difficult to know for certain if a specific cell initiated apoptosis. Also, caspases are used for multiple cell processes, and so their presence may not be a reliable marker for apoptosis. Supporters of anastasis point out that the other appearances of caspases are well defined in development, and their appearance at other, more random times might support the presence of anastasis. All agree that more work is needed to come to a final conclusion.

Other forms of cell death besides apoptosis may be reversible as well. In necroptosis, which is associated with trauma and stress, a protein called mixed lineage kinase-like protein causes holes in the plasma membrane. Those holes can be limited by the ESCRT III complex (Gong et al. 2017). Ferroptosis cannot be stopped by removing the initiating stress, but it can be if lipophilic antioxidants or chelators are also introduced (Tang and Tang, 2019). In entosis, the engulfment of one cell by another, the engulfed cell is usually killed by autophagy and dismantled for recycling by lysosomes, but sometimes it survives and might even reproduce (Hamann et al. 2017). If it does reproduce, the result can sometimes be aneuploidy (extra chromosomes).

Metamorphosis

> As Gregor Samsa awoke one morning from uneasy dreams he found
> himself transformed in his bed into a gigantic insect.
> —Franz Kafka, *The Metamorphosis*

In Kafka's famous novella, Gregor Samsa found that he had changed overnight into an insect. It's a great story, but of course, that kind of transformation occurs only in fiction. Yet other transformations that actually occur in nature are so dramatic that they could also be fiction. A couple of cases are well known even to schoolchildren, but no less amazing for their familiarity. Once they hatch from their egg, caterpillars eat voraciously to grow much larger before surrounding themselves in a pupa. After a period of time, the pupa breaks open, and a beautiful butterfly emerges. In another familiar case, tadpoles gradually lose their tails and gain legs to become adult frogs.

Metamorphosis occurs in several types of organisms, including insects, frogs and toads, fish, and newts and salamanders, and a few others (Holstein and Laudet 2014). PCD is a critical mechanism in metamorphosis. In each transformation, specific cells are programmed to die, and other cells divide and expand to construct new tissues. These changes are controlled genetically and hormonally.

In the cnidarian *Hydractinia echinate*, eggs hatch to release free-swimming planula larvae. The larvae later undergo metamorphosis to become sessile (stationary) polyps. During that process, many cells are lost, and those cells display features associated with apoptosis, such as DNA fragmentation, nuclear condensation, and the presence of caspases (Wittig et al. 2011).

The fish that undergo metamorphosis are members of the infraclass Teleostei. This group, known as the teleosts, represents an enormous number of vertebrate species—over 23,000 (McMenamin and Parichy 2013), and include cod, pike, salmon, trout, catfish, herring, and eels. The changes that occur during their development are just as dramatic as those in frogs. The greatest changes occur when the animals are transitioning from the larval to the juvenile stage. Adult fins form. The fin rays ossify (become bony). The internal organs and sensory systems mature. Scales form. The colors and the body proportions change. In these fish, thyroid hormone is one of the major regulators of metamorphosis. They also have proteins similar to those involved in apoptosis in mammals, with a death domain (Sakamaki et al. 2007).

Insects

Metamorphosis is common in insects. Once insects hatch from their eggs, they begin the process of metamorphosis along one of two paths. Beetles, flies, ants, butterflies, and moths undergo what's called complete metamorphosis; crickets, grasshoppers, and dragonflies undergo incomplete metamorphosis.

Complete metamorphosis has four stages: egg, larva, pupa, and adult (Tettamanti and Casartelli 2019). For example, a butterfly egg hatches to release a larva that we call a caterpillar. Larvae eat a lot. In fact, they consume several times their own body weight each day. And they grow rapidly. In some cases, the larvae develop additional body segments (or instars) as they grow. After a period of growth, the larvae surround themselves with a hard shell, called a pupa. Inside the pupa, a series of developmental steps transforms the larva into the adult form. The caterpillar becomes a butterfly. In contrast to the four-stage development of the butterfly, incomplete metamorphosis has only three stages: egg, nymph, and adult. The nymph is just a smaller version of the adult.

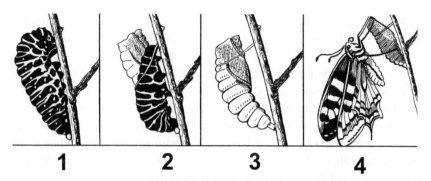

Figure 8.1 Metamorphosis in butterflies. (1) Larva or caterpillar, (2) pupa, (3) cocoon, and (4) adult. Original drawing by Pearson Scott Foresman.

Cell death is a critical component of the activities involved in complete metamorphosis, especially in the transformation of the larva into an adult. Nearly 80 percent of insects go through some form of complete metamorphosis, but the extent of PCD differs from species to species (Rolff, Johnston, and Reynolds 2019). PCD in metamorphosis is by autophagy or apoptosis. Organisms use PCD to remodel cells and to recycle the cell contents. Apoptosis involves other phagocytic cells to recycle the components of the dead cell, but autophagy is accomplished by the cell itself. Apoptosis usually occurs earlier than autophagy, and hormones are intimately involved in regulation of metamorphosis.

The moth *Manduca sexta* has been the subject of many studies of metamorphosis. In this insect, metamorphosis involves cell proliferation, modification of existing structures, and PCD. For example, a large part of the central nervous system is replaced. The cell deaths are scattered, but the muscles are almost completely new. In fact, about 40 percent of the intersegmental muscles are lost even before the pupa is formed. The fate of the muscles differs. Intersegmental muscles are not eliminated by phagocytes or neighboring cells, but mesothoracic muscle fibers are. Also, the latter are not completely degraded; they are instead used to build new structures.

Neural tissue is lost during metamorphosis as well. Motor neurons have been carefully studied, perhaps because they are the easiest to identify and track. The motor neurons that are lost are generally those for which their target is also lost. In other words, those segments that are no longer needed in the adult are lost, and the neurons that connect to them are also subjected to PCD. Sex differences occur; for example, the imaginal midline neurons that innervate the sperm ducts are absent in females, and those that innervate the oviduct are absent in males. Neurons in the brain die by PCD as well.

Most gene expression during pupation remains the same, but a few candidates for death-associated genes have been identified, and during pupation their expression increases (possible mechanisms for controlling gene expression include decreasing transcription or increasing RNA degradation; microRNAs might also be involved in regulating gene expression at the translational level). Among the genes identified as being increased in expression were those for ubiquitin. Since phagocytosis is not a significant factor in the moth pupal cells, perhaps the degradation is accomplished by the ubiquitin-proteasome system. Apolipoprotein III is greatly enhanced as well; it's speculated that this could be helpful in transporting lipids, but the real function of the protein is not clear.

Amphibians

Metamorphosis in amphibians is every bit as dramatic as in insects. Frog eggs hatch to release tadpoles, the larval stage of a frog. Most tadpoles live in the water and breathe through gills. They lack legs and swim using a tail that is nearly twice the length of their body. Transformation of a tadpole into an adult frog involves the exchange of larval organs for adult organs and a change from an aquatic existence to a terrestrial one. The key to the transformation is thyroid hormone.

The changes begin with the gradual appearance of the back and then front legs and the slow disappearance of the tail. Lungs begin to replace the gills. The mouth grows much larger, and the intestines are transformed into ones that can handle a carnivorous diet (Ishizuya-Oka, Hasebe, and Shi 2010).

In the African clawed frog (*Xenopus laevis*), a favorite experimental model system, metamorphosis takes three days and can be divided into three periods: premetamorphosis, prometamorphosis, and climax (Nakajima, Fujimoto, and Yaoita 2005; Exbrayat et al. 2012). Premetamorphosis begins when the egg hatches and lasts until the thyroid hormones T3 and T4 begin to be produced, initiating the next phase. During prometamorphosis, the animal undergoes multiple changes, including the development of the hind legs. At the beginning of this stage, apoptotic cells appear in the spinal cord and caudal spinal ganglia. PCD continues in the spinal cord, limb motor neurons, ependymal cells (neural support cells), and more. The optic nerve remodels as the eyes reposition to facilitate binocular vision. During the climax phase, levels of thyroid hormone reach a peak, and many changes occur: the front limbs appear, the head shrinks, the gills disappear, most of the internal organs are replaced, and the tail is absorbed over a few days. Several PCD processes are involved in these changes, including apoptosis, autophagy, and a modified apoptosis that does not involve lysosomes. The epidermis of the tail shows classic apoptosis; the muscle fibers show many of the features of apoptosis except that the apoptotic bodies are engulfed by macrophages. Proteases are also induced and attack the extracellular matrix; as a result, some cells die by anoikis. Matrix metalloproteinase 9 (MMP-9) causes apoptosis in the larval epithelium by degrading the ECM and is also involved in intestinal remodeling. Apoptosis is reported in the metamorphosis of the digestive tract in two other frog species, the northern leopard frog (*Rana pipiens*) and the Argentine horned frog (*Ceratophrys ornate*) (Cruz Reyes and Tata 1995). The intestine makes a major change as the organism goes from omnivorous tadpole to carnivorous frog—the simple larval tube intestine is replaced by a more complex adult intestine of multiple folds, connective tissue, and muscle. The elimination of the larva intestine is regulated by thyroid hormone and involves apoptosis and the activation of caspases.

The European eel (*Anguilla anguilla*) has a complicated life cycle that includes metamorphosis and extensive physical changes. Eels hatch from eggs as small transparent leaf-shaped larvae in the Caribbean. For one to three years they stay at sea, riding the Gulf Stream across the Atlantic Ocean. By the end of the trip they have metamorphosed into small transparent forms with a more eel-like shape, at which point they are known as glass eels. In Europe, their pigmentation develops, and they become known as elvers. They make their way up the freshwater rivers and streams to the headwaters, where they live for many years in a sexually immature adult form (in which they are known as yellow eels). After perhaps ten to fifteen years, they become sexually mature (at which point they are known as silver eels), and they make their way back down the rivers to the sea and again cross the Atlantic to breed. Interestingly, one of the long-standing mysteries about eels was why they seemed to lack testes. In fact, they have none until they arrive back in the Caribbean to spawn.

The eels experience two metamorphoses: the first is from the larval stage to the juvenile stage, and the second is from the juvenile stage to the adult stage. Pujolar et al. (2015) examined patterns of single nucleotide polymorphisms and found a set that seemed to be enriched. These were found in genes that encoded for signal transduction pathways, including MAPK/Erk, calcium, and gonadotropin-releasing hormone signaling. Trautner et al. (2017) looked at the possibility that epigenetics might be involved in these changes. They examined the patterns of methylated DNA in glass eels from the British coast and yellow and silver eels from the Rhine River. They found only minor differences. Differences in the methylation patterns in the gills might be involved in the changes from salt to fresh and back to salt water. Perhaps more interestingly, differences in brain patterns might be involved in the metamorphoses.

9

Programmed Cell Death in Humans

Despite Franz Kafka's brilliant imaginings in *The Metamorphosis*, humans do not undergo metamorphosis. Nevertheless, programmed cell death is critical for our development and physiology; during our development from an embryo to an adult, specific cells live, and others die (Fuchs and Steller 2011).

As we saw in Chapter 8, all across the animal kingdom death shapes individual organisms during development. As human embryos develop, excess cells are made, and then the extra cells are eliminated by apoptosis as the tissues and organs are completed. Later on, many cells have a normal life span, after which they simply wear out, die, and are discarded. Some cells, especially those exposed to the external world, wear out rapidly with that exposure and must be replaced. These include skin cells, the cells that line the intestines, and blood cells.

Development

Oocytes

During development, every human female begins with about 7 million oocytes, or egg cells, but by the time of birth she retains only one-third of those, which are in what's known as ovarian primordial follicles (Vaskivuo and Tapanainen 2002). Once the woman reaches maturity, a number of those follicles begin to grow at each menstrual cycle, but typically only one releases an oocyte during ovulation. The rest of the oocytes is culled by apoptosis; the Bcl-2 family of proteins regulates apoptosis in oocytes, and mitochondria are also involved. Furthermore, in the follicles that do not release an oocyte in that cycle, cells called granulosa cells must be eliminated by apoptosis. The follicle that ovulates then forms the corpus luteum, a mass of cells that will produce progesterone to maintain the uterine lining, or endometrium, in early pregnancy. If a pregnancy does not occur, the corpus luteum must be eliminated so that the cycle can begin once more, and again, apoptosis is the mechanism that makes this happen. Caspase-3 seems to be involved in the regression of the corpus luteum if a pregnancy does not occur.

Mitochondrial DNA is inherited through the mother, and thus, it would be beneficial for any follicle with damaged DNA to be eliminated.

The Biology of Death. Gary C. Howard, Oxford University Press. © Oxford University Press 2021.
DOI: 10.1093/oso/9780190687724.003.0009

Fingers and Toes

As discussed briefly in Chapter 8, the development of fingers and toes is a classic example of how apoptosis is involved in determining human body shape (Hernandez-Martinez and Covarrubias 2011). The hands begin their development as flattened hand plates at the distal ends of the growing arms (Brill et al. 1999; McCarthy 2003). At the tip of each digit, the ectodermal ridges induce a type of embryonic tissue called mesenchyme to begin to form phalanges, and cartilage begins to develop. Bone morphogenic proteins kick in; they have two different types of effect in the developing hand, inducing both growth and apoptosis. The extracellular matrix acts as a scaffold for cell aggregation and may influence the effects of growth factors. As the digits begin to form, they are still webbed. Retinoic-acid-induced cell death (apoptosis) is responsible for removing the extra tissue, "sculpting" the hand and allowing the fingers to appear as individual structures (Suzanne and Steller 2013). A similar process occurs with the toes on the feet. This is only one example of the involvement of apoptosis in the organization of cells into the final body form. Not all organisms undergo this process; ducks and bats, for example, lack cell death in the extremities, and so ducks have webbed feet and bats have webbed "hands" (which we call their wings). Nor is the process always successful in humans. For example, in the condition known as syndactyly, the fingers do not separate properly and remain joined by tissue. Studies in mice have shown that this condition results from a defect in apoptosis (Schatz, Langer, and Ben-Arie 2014).

Neural Development

Neurons are different from other cells. They must form multiple (perhaps thousands of) connections to other neurons. Those connections store our memories and facilitate our ability to think and reason. Furthermore, connections are continually being made and replaced as our experiences continue and change. Yet many neurons die during development. As a human embryo develops, many more neurons are created than are needed, and more than 50 percent of them die as nonfunctional circuits and excess neurons are eliminated. This great loss of cells ensures that proper connections are made in the central and peripheral nervous systems.

PCD is an important factor in the developing nervous system. Many mechanisms are involved, including intrinsic and extrinsic apoptosis, oncosis, necroptosis, parthanatos, ferroptosis, sarmoptosis, autophagic cell death, autosis, autolysis, paraptosis, pyroptosis, phagoptosis, and mitochondrial permeability transition (Fricker et al. 2018). Apoptosis regulates many aspects of the nervous system, including the sexual differentiation of the medial preoptic nucleus, development of the lining of the nose, and the optic nerve from the back of the eyes to the brain, and the nerves that connect of central nervous system to other parts of the body. The nerve growth factor and its receptor are important for the survival or death of the neurons in the retina of our eyes and oligo-dendrocytes, which add the insulating myelin layer to nerves. The receptor binds to nerve growth factor strongly or weakly. If it binds strongly, it will act to encourage the survival of the cell. If it binds weakly, it will induce PCD in the cell.

Neurons are postmitotic—that is, they have exited the cell cycle and no longer divide—so they must live essentially for the lifetime of the organism in order for those connections to be maintained. One of the hallmarks of most neurodegenerative diseases is the loss of neurons. The type of neuron lost is fairly specific for each disease. For example, in Parkinson's disease, dopaminergic neurons die. In Alzheimer's disease, it is neurons in the hippocampus. Some theories attribute neuronal death in these diseases to PCD, while others have invoked autophagy.

Immune System

As we have seen, the immune system is a critical element of the body's ability to defend itself from pathogens and other invaders. Once a pathogen is detected, the immune system responds with cells and antibodies. It is estimated that between 50 and 200 T cell precursors for a specific antigen each expand10,000- to 50,000-fold in response to a pathogen (Hedrick, Ch'en, and Alves 2010). However, when the danger is past, it is important for the system to go back to a homeostatic condition, so the additional cells must be removed efficiently (Krammer, Arnold, and Lavrik 2007).

While the immune system must rapidly detect and respond to pathogens, it is equally important that it not respond to the host's own tissues and proteins, as discussed briefly in Chapter 8. How can the immune system make this critical distinction? The immune system "learns" to tell the difference through clonal selection, a mechanism suggested by Frank M. Burnet and later proved by Peter Medawar. The diversity of the T cell receptors of our immune system is immense; information for an immune response to pretty much everything is encoded in our genes (which contain the information for many more responses than originally thought because, as we now know, the genes rearrange themselves). When an antigen enters the body, it activates the receptor on a specific T cell; that cell then clones itself to produce sufficient amounts of antibodies against that antigen. Among those many T cell receptors are numerous ones for our own bodies; those cells have to be eliminated so that they don't multiply and produce antibodies against the self. During maturation of the immune system, those cells that would react against the self are eliminated by apoptosis in the thymus (Medzhitov and Janeway 2002). We have discovered that stimulation of T cells by specific molecules, such as anti-CD3 antibodies or TCR antigen, induce apoptosis in the T lymphocytes developing in the thymus. Very few T lymphocytes make it through this rigorous process; in fact, only about 3 percent survive. Not all apoptosis begins with external signals. Even in cells lacking any kind of death receptor, apoptosis still occurs, so there has to be an intrinsic pathway as well. The Bcl-2 family was found to be involved. Immune pathways are also sex-associated (Ahnstedt et al. 2020). Aged males have more problems with sensory and motor nerves than females, but they also have a greater brain T-cell immune responses. PCD is important for the immune system even after development and maturation. Natural killer cells and cytotoxic T lymphocytes use it to kill viruses and tumor cells. They release cytotoxic granules that contain a protein called perforin, which causes pores to form in the attacked cells or viruses, and proteases called granzymes, which enter through the pores and initiates apoptosis. Alternatively, gasdermin B can be released to enter the cells and cause pyroptosis, an inflammatory variation on apoptosis (Zhou et al. 2020).

The cells of the adaptive immune system must be ready to divide rapidly to mount an effective defense against pathogens. Just as importantly, those cells must return to their normal numbers after the pathogen has been eliminated. Apoptosis seems to have a significant role in reducing the number of immune cells after an immune response, though the mechanism is not completely clear; extrinsic triggers are suspected. Necrosis is a possibility, too—the addition of necrosis inhibitors to dying cell cultures stopped the dying. It seems likely that apoptosis, autophagy, and necrosis may all play a role, depending on the infectious agent. We do know that autophagy is involved in the development of the adaptive immune system; it assists in thymocyte development and maintains homeostasis in peripheral T and B lymphocytes. The protein ATG7 is required for the maintenance of mouse hematopoietic stem cells, which produce these lymphocytes (Mortensen et al. 2011); without that protein, the cells proliferate and the mice die within weeks (those proliferated cells also showed accumulations of mitochondria and damaged DNA). Autophagy is a key death pathway for T cells that lack FADD or caspase-8. Autophagy is also used to shape the repertoire of T cells and is critical for ensuring that the immune system can differentiate between self and non-self (Nedjic et al. 2008).

Bones

In the embryo, cartilage develops first, laid down by specialized cells called chondrocytes, which die by apoptosis after completing their work. In the process of ossification, or the development of bone, minerals come to surround the cartilage in the appropriate places (Boskey and Coleman 2010). A growth plate at each end of the bone controls the development of the bone.

Although we generally see bone as a permanent structure, it is, in fact, a dynamic tissue: old bone is continually reabsorbed and new bone is synthesized. Two types of cells are involved in this process. Osteoclasts destroy existing bone material, and osteoblasts synthesize new bone. Clearly, a careful balance is critical, and apoptosis is used to maintain that balance. Osteoclasts in particular work very rapidly, and it is important to control their activity. Studies of these cells are complicated by the very small number of osteoclasts compared to other bone cell types, but apoptosis seems to be a common form of death for them. In general, factors that cause bone resorption also inhibit apoptosis and vice versa, which can lead to the bones thinning out and becoming more vulnerable to breakage (osteoporosis). Other factors stimulate osteoclast apoptosis and thus reduce the rate at which bones thin. Estrogen is one substance known to encourage osteoclast apoptosis; this might be the reason hormone replacement therapy can be effective at preventing osteoporosis in postmenopausal women.

Red Blood Cells

Erythrocytes, or red blood cells, which carry oxygen in our blood, are among the most common cells in our body. They wear out after about 120 days. Thus, every day a very large number of these cells must be disposed of. Apoptosis is a common mechanism for dealing with damaged and worn-out nucleated cells, but erythrocytes lack nuclei and mitochondria. Nevertheless, dying erythrocytes exhibit several of the classic features of apoptosis, and the term *eryptosis* has been suggested to describe this form of PCD (Föller, Huber, and Lang 2008; Pretorius, du Plooy, and Bester 2016). Eryptosis is a complicated process that involves ion channels and multiple signaling molecules. Many triggers of eryptosis have been suggested, including the binding of specific antigens, such as glycophorin-C, thrombospondin-1 receptor CD47, and CD95/Fas. Eryptosis has a low threshold and occurs in many diseases and situations. Ca^{+2} ions might also be involved. They might cause the cell membrane to form vesicles and to rearrange so that phosphatidylserine molecules are exposed at the cell surface. These ions might also stimulate the cysteine endopeptidase calpain, leading to loss of the cytoskeleton and blebbing of the cell membrane.

Skin Cells

The bodies of land animals, including humans, are exposed to the environment and its significant stresses, including desiccation, mechanical damage, and infectious agents (Costanzo et al. 2015). Arthropods have a strong exoskeleton to protect them. Tetrapods (e.g., reptiles, amphibians, mammals, and birds) have skin.

The skin is the largest organ in the human body. It has three layers. The outer layer of skin is composed of a layer of dead cells and extracellular material that is continually renewed by dividing keratinocytes. In the bottom layer of the epidermis are melanocytes, which produce the pigment (melanin) that gives the skin its color. The next layer, the dermis, is made up of connective tissue, blood vessels, and touch, pressure, and pain sensors. It also contains hair follicles and sweat glands. The deepest layer, the hypodermis, contains fat and connective tissue.

Homeostasis in the skin involves the differentiation of keratinocytes from epidermal stem cells, a type of apoptosis called cornification (mentioned briefly in Chapter 8), and finally their removal by desquamation. Cornification also produces other skin structures, such as nails, hair shafts, and the papillae of the tongue. Cornification differs from typical apoptosis, but some processes are similar. Although cornification involves the death of cells, those cells are not discarded, but used as building blocks to form the protective outer layer of the skin. It is important that the keratinocytes not die prematurely and initiate various inflammatory and other processes that would interfere with the construction of the cornified skin layer. Therefore, the cells have several protective mechanisms that protect them from unwanted apoptosis or necrosis. First, the cellular organelles and other materials are degraded by caspases and replaced by a cytoskeleton. Proteins are cross-linked by transglutamination at the cell periphery to form a cell envelope. Finally, the corneocytes are formed into a strongly connected but dead structure.

Intestinal Cells

The intestinal wall forms a critical barrier that prevents waste and microbes from crossing the intestinal wall and entering the abdominal cavity, but still allows the absorption of nutrients and water. The lining of the intestines comprises a very large area that is covered with intestinal epithelial cells. The surface area of the intestines is even greater because of its structure: protrusions from the intestine into the lumen or open cavity of the intestines are called villi, and the depths near their bases are called crypts.

With the continual exposure to food, waste, and bacteria, the barrier cells wear out over time and must be replaced. This exchange requires a careful balance of the production of new cells and the shedding of old cells. Stem cells near the crypt continually produce new cells that migrate up the villus. PCD occurs at two sites: in the crypts (where it is called spontaneous apoptosis) and at the villus tip (where the process is called anoikis). As we've seen in Chapter 8, anoikis occurs in anchorage-dependent cells like the intestinal epithelial cells—they sense where they are in the intestinal matrix by interacting with the extracellular matrix and the cells near them. When they begin to become detached, they sense that the new input they are getting (e.g., from integrin) is not appropriate, and so they initiate PCD (Gilmore 2005).

Why cells are shed at the villus tip is not clear. It might be a simple case of crowding at the tip. As more cells move up the villus, the space at the tip becomes packed with cells, and some must be detached and lost. Another theory implicates inflammation. As apoptosis is initiated, some extracellular vesicles and chemokines may attract immune cells. The microbiota of the intestines also play a role in inducing apoptosis, with certain groups of organisms (e.g., *Salmonella*, *Shigella*, pathogenic *Escherichia coli*, *Helicobacter pylori*, and *Cryptosporidium parvum*) enhancing apoptosis.

Disease and Injury

Apoptosis is a key part of neuronal injury and disease (Mazarakis, Edwards, and Mehmet 1997). Caspase activity has been found in most studies of postmortem brain tissue from Alzheimer's and Parkinson's patients, indicating apoptosis. What causes the loss of neurons in Parkinson's is unknown, but the leading candidates are toxins, oxidative stress, mitochondrial defects, aggregated proteins, and a faulty ubiquitin-proteasome system. Apoptosis has been implicated in Huntington's disease, ALS, spinal muscular atrophy, ataxia telangiectasia, and retinitis pigmentosa. Developmental delays in Down's syndrome have also been linked to inappropriate apoptosis. In Parkinson's, dopaminergic neurons in the substantia nigra die by apoptosis. In Alzheimer's, accumulation of beta-amyloid protein can induce neurons to undergo apoptosis.

Infections

Cell-mediated cytotoxicity is an important mechanism for protecting against intracellular pathogens. These are pathogens that have eluded our other lines of defense and have reached and infected a cell. The infected cells "ask" for help by breaking down some of the pathogen's proteins and presenting those protein fragments on their surface. CD8 cytolytic T lymphocytes (CTLs) recognize those antigens and use two pathways to kill infected cells. First, they secrete cytotoxins that cause cell death by necrosis. Second, they secrete proteases that interact with other factors to induce apoptosis by activating caspase-3.

Cell-mediated cytotoxicity also works against tumor cells. Antibodies bind to antigens on the surface of tumor cells. Natural killer (NK) cells with CD16 bind to those antigens, releasing enzymes that destroy the tumor cells. Other immune cells, such as macrophages, neutrophils, and eosinophils, might also be involved. NK cells lack antigen-specific receptors. However, they can interact with specific antibodies that have bound the antigens. Once bound, the NK cells release cytotoxic granules into the intercellular space between the NK cell and the tumor cell. The granules contain perforin, which causes holes to form in the cell membrane of the cancer cell and lead to the death of the cell.

Autophagy can be used by cells as an arm of the immune system. Using autophagy, cells can capture intracellular bacteria and destroy them (Puleston and Simon 2013). The degraded fragments of the pathogens can then be recognized by the innate and adaptive immune systems.

In HIV infection, only a few cells seem to be infected, but many other seemingly uninfected cells die. This is called "bystander killing," and it was one of the long-standing mysteries of HIV infection. That mystery was solved by Doitsh et al. (2014). They showed that only a few cells were killed by caspase-3-mediated apoptosis. The great majority of the cells (the bystanders) were killed by caspase-1-mediated pyroptosis during an abortive HIV infection. As mentioned in Chapter 8, pyroptosis is an intensely inflammatory form of programmed cell death in which the cell bursts open and the cytoplasmic contents and proinflammatory cytokines, including interleukin-1-beta (IL-1B), are released. The result is the death of CD4 T cells and chronic inflammation. A vicious cycle results in which the released cytokines recruit additional T cells to the killing site.

Autophagy is a key antimicrobial defense, but no biological system is perfect. As is often the case, pathogens have found ways to exploit it (Orvedahl and Levine 2009). Bacteria that prevent the fusion of autophagosomes to lysosomes can escape destruction. Autophagosomes are cell organelles that contain material that is meant to be degraded once the autophagosome merges with a lysosome. Some bacteria, such as *Listeria monocytogenes*, even use the material in the autophagosome as a source of raw materials. Some viruses have evolved ways to stop autophagy signals.

Apoptosis and Cancer

Apoptosis is also a mechanism to protect against cancers. While most of the body's cells stop dividing once the tissue or organ has achieved full growth, some reenter the cell cycle and "forget" the limits on their growth. They pose a threat to the rest of the body if they survive. This reentering of the cell cycle can be caused by DNA damage or certain viruses. Cells with these injuries can become precancerous. The apoptotic system recognizes precancerous cells and even cells with DNA damage. Proto-oncogenes (e.g., *c-fos*, *c-jun*, and *c-myc*) trigger apoptosis. If these cells undergo apoptosis, the threat to the rest of the organism is removed. Apoptosis is usually initiated by the cell's own internal mechanisms; however, sometimes this system fails. In that case, the immune system may detect the damaged cell and trigger apoptosis through external signals. In recent years, there has been considerable research into ways that these pathways might be manipulated therapeutically to treat cancer.

10

Death in Plants

> There is a Reaper, whose name is Death,
> And with his sickle keen,
> He reaps the bearded grain at a breath,
> And the flowers that grow between.
>
> —Henry Wadsworth Longfellow

Plants have often been used as an analogy for the cycle of life. Tender green shoots sprout in the early spring, and the plant begins its growth to maturity. It flowers in the spring. The flowers fall, and fruits mature through the summer. Autumn brings the harvest, and the leaves change colors and drop. In winter, the cycle of life ends with death. These images are very familiar to us all, but what is not so familiar is how death is actually incorporated into many parts of this image. Programmed cell death is the mechanism that causes the leaves to change colors to red, orange, and gold and fruit to fall.

Plant life spans vary even more than those of animals. Some plants live only one season, but others live for very long times. A bristlecone pine (*Pinus longeval*) in the White Mountains in eastern California is over 4,800 years old. A colony of quaking aspens (*Populus tremuloides*) in south-central Utah is estimated to be 80,000 years old; the trunks of those trees are interconnected through an extensive underground root system, so they are essentially a single clonal organism. Over the years, long-lived plants are subjected to a lot of wear and tear, including in their chromosomes. One might think mutations would collect over time and eventually affect reproduction and other activities. However, this does not seem to be the case. Examination of the bristlecone pine showed few mutations and little effect on reproduction.

Cell Death

Although some plants live for long times, they are not immortal. At some point, like animals, plants succumb to an insult or insults and ultimately die. Plants die from many causes, including drought, heat, cold, fires, and diseases. Plants

The Biology of Death. Gary C. Howard, Oxford University Press. © Oxford University Press 2021.
DOI: 10.1093/oso/9780190687724.003.0010

vary in their resistance to these causes. Natural selection allows some species to thrive and others to die, and sometimes even species that have thrived for long periods of time lose out if environmental conditions change too quickly for them to adapt. Even today, these changes are taking place. For example, climate change is already having an effect on plants. Plant species are changing their ranges as the planet warms. If they cannot adapt quickly enough so that they can grow and compete successfully with other plant species, they may go extinct. Plants also experience aging and senescence. And, like animals, they have evolved mechanisms that take advantage of death to assist them in various aspects of their lives.

Differentiating living and nonliving is somewhat more problematic for plants than for animals. Most animals are motile for at least some part of their lives, and so if they are still moving, it's easy to say they are alive. Plants are generally rooted in one place and do not move. So that obvious test of life or death is out. Plants also lack a heartbeat, and their respiration has no easily recognized signal, such as breathing. Even an obviously living plant will not fog a mirror held close to it.

The very structure of plants complicates the recognition of living and dead. Trees are a great example. Some parts are clearly alive (e.g., leaves, leaf and flower buds, roots, and the cambium, the layer of cells just under the bark). Other parts have ceased metabolism and lost their cytoplasm but nevertheless are critical for the health and growth of the tree. The bark provides a strong covering to protect against trauma and pest infestations. The inner wood provides support to allow the tree to grow tall and channels for water and nutrients to move up and down the tree. But are those parts of the tree alive or not? When we see a tree branch on the tree, we see it as alive, but when it is lying on the forest floor, we call it dead. One way to look at it is that trees are mostly not quite alive. Some estimates, in fact, are that only about 1 percent of a mature tree is alive.

Senescence, as we have seen, is the loss of biological function associated with aging. Two general models have been suggested to explain senescence in plants. The first model involves a "death hormone." To be sure, hormones are deeply involved in the regulation of senescence in plants; the hormones that tend to promote senescence include abscisic acid, ethylene, jasmonic acid, salicylic acid, gibberellins, brassinosteroids, and strigolactone. These hormones work in opposition to the cytokinins, which tend to promote growth and development. While at this point no unique "death hormone" has been identified that invariably leads to senescence and death, this model remains at least a possible mechanism.

The second model involves nutrients. Plants carefully balance their investments of resources and continually move nutrients from place to place within the plant, depending on the need. One might imagine the plant reallocating resources from dying areas of the plant to those that are doing better. The two most important nutrients are fixed nitrogen and sugars (e.g., sucrose, the hexoses, glucose, fructose). Different sensors detect the two groups and determine their movement throughout the plant. Sucrose is broken down into glucose and fructose at parts of the plant called "sources" so that the sugars can be transported to areas of need, called "sinks" (Thomas 2013). Genes are regulated to express the appropriate enzymes that allow this process. The balance between sources and sinks is critical, and it is maintained by the opposing actions of the sugar sensors and cytokinins. Cytokinins encourage growth, but the movement of sugars (such as from leaves to seeds) signals senescence (in this case, of the leaves). Strong sinks have more cytokinins that attract sugars from sources.

Death ends life and the biological processes in plants, but organic reactions continue. The term *acherontic* has been suggested for the transition between life and death, when the molecules, such as proteins, lipids, chlorophyll and other pigments, are being withdrawn from a dying leaf or other part and distributed to other living parts of the plant (Thomas 2003) (the term comes from Greek mythology, where Charon would row dead souls across the river Acheron so they could enter the underworld). The processes of death can be at different stages in the same leaf. Senescence begins at the leaf margin and works its way back to the veins and stem. Cytokinins—plant hormones that prompt cell division, plant growth, and the development of fruits and seeds—produced in the roots keep the leaves working. The cytokinins are redirected from the leaves to the seeds to ensure that the seeds mature properly. Without the cytokinins, the leaves begin to age.

Plants are divided into annuals or perennials, according to the survival strategy that they use. Death in plants is linked to reproduction. Annual plants (monocarps) undergo a single round of reproduction and then die. This process has been called "big bang" senescence. Perennials (polycarps) typically spend the first year (or more) in growth and development; after they reproduce, they die soon afterward. Annuals commit all of their resources to seeds for the next generation. But perennials have a head start over seeds in that they already have part of their vegetative structure ready to go in the spring. In each case, the plant selects parts for retention and others for death. For annuals, all parts of the plant are discarded except for the seeds. In perennials, select parts (e.g., some stems, roots) are retained, and the remainder are pruned through programmed cell death.

Annuals regulate the use of the nutrients that they receive and direct them to maximize reproduction at the expense of later growth. For example, growing leaves synthesize the CO_2-fixing enzyme rubisco in the green leaves of a plant. Once the leaves are fully grown, the synthesis stops, and almost immediately the degradation of the enzyme begins. The components of the degraded rubisco are then recycled for use in the plant. Similarly, before a tree's leaves fall in the autumn, proteins in the leaves are recycled into the bark to preserve the valuable nitrogen.

Programmed Cell Death

Like animals, plants have harnessed death to facilitate and ensure living. Plants and animals share a lot of features. Both are eukaryotes and so have cell membranes and membrane-bound organelles. However, they diverged from each other a very long time ago in terms of evolution, and their cells have a number of significant differences. Plant cells contain chloroplasts (structures where photosynthesis occurs), and chlorophyll and other pigments allow plants to use photosynthesis to make food from sunlight and carbon dioxide. Animals cannot make their own food and depend entirely on consuming other animals or plants. Animals also take in oxygen and release carbon dioxide, but plants use carbon dioxide while they are photosynthesizing. Finally, animals have well-developed senses, but plants have few. Most importantly, plant cells also have a rigid cell wall and often large vacuoles. Vacuoles contain water and other material in a large membrane-enclosed sac called a tonoplast. These structures cause the various types of death to be different in plants than they are in animals. Those differences complicate the study of programmed cell death in plants.

Plants use multiple types of PCD that are deeply intertwined with normal growth and development and with disease, senescence, and trauma. The major forms have the same names as in animals: apoptosis, autophagy, and necrosis. However, they are not exactly the same because of the distinct differences between plant and animal cells. As we've seen, while both types of cells are bounded by a plasma membrane, plants also have a rigid cell wall that obviates some features of apoptosis. These differences have resulted in some controversy about whether plants actually exhibit apoptosis. However, many scientists agree that plants show sufficient apoptosis-like actions for the process to be considered a form of apoptosis.

The study of PCD in plants has also been hindered by the fact that in a typical plant, dying tissue is only a very small percentage of the healthy tissue. Thus, some scientists have decided to examine PCD in cell culture, where the conditions can be better controlled. Cultures are uniform, the cells are quite accessible, and they are less complex than whole plants with multiple differentiated tissues.

Plants do undergo senescence, and some of its features are similar to those of programmed cell death. For example, DNA is cleaved, nuclei are condensed, and the cytoplasm shrinks away from the cell wall.

Apoptosis

Apoptosis is well understood in animal cells (see Chapter 8). To briefly review, animal cells shrink, "blebs" appear on the cell surface, the chromatin condenses, and the DNA fragments. In the end, the cell fragments into smaller membrane-bound packages (apoptotic bodies) that are degraded by phagocytes. In animal cells, mitochondria are believed to be the major regulator of apoptosis. No inflammation results from apoptosis. .

Plant cells, however, are unable to display some of the classic characteristics of apoptosis in animal cells. Van Doorn et al. (2011) pointed out that the rigid cell walls in plants do not allow the cells to break down into apoptotic bodies; instead, when plant cells shrink, the cell membrane is damaged and leaks. Also, plants lack phagocytic cells that engulf and degrade bacteria, small cells, or cell fragments.

Some of those who insist on a strict definition of apoptosis prefer the term *vacuolar cell death* for the aspects of plant death that resemble apoptosis in animals. While plant cells lack lysosomes that are important in animal cell death, they do contain large vacuoles. Some of those vacuoles contain lytic enzymes that perform degradative functions similar to those of lysozymes in a process that resembles autophagy: actin filaments form, the nuclear membrane segments, and organelles are unchanged until the tonoplast fails. Once that membrane fails, the cell is cleared out quickly. As this process proceeds, the cytoplasm decreases and the lytic vacuoles increase in size. The lytic vacuole folds back on itself to create a pouch, and its membrane fuses with that of the target organelle or cell fragment; the contents of the organelle or cell fragment are then degraded. As a final step in vacuolar death, the enlarged lytic vacuole bursts and releases its enzymes into the cell, where they continue to degrade the remaining parts of the cell. The process might take several days, and the cell wall is sometimes degraded as well. This type of cell death occurs in embryonic development, in organ and tissue formation, and in the formation of pollen and ovaries. The genes responsible for vacuolar death are not known, though the apoptosis gene ATG has been implicated in cell death in *Arabidopsis*, a member of the mustard family that is often used as a model organism in plant biology studies.

Autophagy

Autophagy is also involved in multiple important processes in plant cells, such as seedling establishment, plant development, stress resistance, metabolism, movement of materials within the cell, and reproduction (Michaeli et al. 2015). It allows plants to degrade cells and to selectively degrade specific compartments and protein complexes, thus regulating cellular processes. Its role in recycling plant components is critical: plants have to manage their energy stores carefully, and in times of nutrient shortages, autophagy degrades excess components so they can be transferred to areas of more critical need.

Two key signaling pathways are involved in autophagy: mTOR and sucrose nonfermenting 1-related kinase 1. Autophagy can be extraordinarily selective in its targets as a mechanism for regulating growth. Autophagic cell death is a common feature of plant development. A large vacuole forms and ruptures to release hydrolytic enzymes that degrade the cell contents. Autophagy may be the most important type of PCD in plants, especially where there is time for resources to be transferred from one part of a plant to another. Autophagy does not involve chromatin condensation, but it does involve vacuolization of the cytoplasm.

Necrosis

The third type of plant cell death is necrosis, which results from the localized death of cells. One of its most common manifestations is black or brown spots on leaves or fruits, but necrosis also occurs in multiple other plant tissues and has many different symptoms. Leaves may curl over or turn dry and leathery. Calluses or galls may form on branches. Roots or branches may grow in large number and closely together. All of these are manifestations of necrosis in different plant tissues. Necrosis often occurs when pathogens infect a plant.

Necrosis is relatively poorly understood. It does not involve the swelling of lytic vacuoles, as in vacuolar cell death. In necrosis, the cell fails to control osmosis, and so water and ion influx continues. It usually involves a gain of cell volume, swelling of some organelles, and the rupture of the plasma membrane. Mitochondria swell. Levels of reactive oxygen and nitrogen species increase, and the output of ATP drops as respiration breaks down. Unlike vacuolar death, necrosis occurs rapidly, on a scale of minutes to a day.

In plants, necrosis and apoptosis have different thresholds for activation. For example, moderate heat stress causes apoptosis-like death, but higher temperatures cause necrosis. In addition, the timing of events differs. Lanthanum chloride blocks apoptosis, but only if it is given before heat shock; this sequence suggests that, like in animals, an influx of calcim occurs early in cell death in plants.

The release of certain proteins (e.g., cytochrome c) from mitochondria is a classic feature of animal apoptosis, and it is also seen in plants. Animals undergoing apoptosis exhibit caspase activity; plants do not have caspase, but they do have a caspase-like enzyme that plays a similar role. And both animals and plants experience DNA degradation and loss of mitochondrial transmembrane potential. Mitochondria may be the central regulator for PCD in plants, as loss of the outer mitochondrial membrane is a key feature of all three types of cell death.

Necrosis in plants has three forms: necroptosis, pyroptosis, and ferroptosis. The plasma membrane ruptures and releases the cellular contents. Some of those contents elicit an immune response.

A variation of necrosis, ferroptosis, requires iron and involves oxidation but not apoptosis (Distéfano et al. 2017). It is induced by heat shock, but not by normal development. As in animal cells, a ferroptosis-like cell death in root cells involves a loss of glutathione and an increase in levels of reactive oxygen species. The researchers examined root cells of *Arabidopsis thaliana* at temperatures of 55° and 77°C. They also used ferrostatins, compounds that reduce levels of iron in the cells. The ferrostatins induced cell death at 55°C. At 77°C, the cell died, but it was by an unregulated mechanism. The parallels between cell death in animals and plants by similar iron-involved mechanisms suggest that this method of programmed cell death has been conserved since the time when plants and animals diverged in evolution. Furthermore, no physiological or developmental function of these types of cell death has been identified. They seem to be limited to defenses against pathology.

Death in Growth and Development

PCD is important for the proper development of all multicellular organisms, including plants. It is a critical mediator of reproductive structures, the remodeling of cells and tissues, and senescence. Apoptosis, autophagy, and necrosis all have roles.

Reproduction

PCD is intimately involved in many phases of reproduction. In seed plants, the ovule has three parts: the integument, the outer protective layer; the nucellus, the inner layers; and the female gametophyte, which produced the egg. Right after fertilization of the egg, the nucellus begins to deteriorate, and the degraded material is then used to feed the embryo. The degradation of the nucellus is thought to be initiated by small structures called ricinosomes that contain proteases and are budded off from the endoplasmic reticulum (Schmid et al. 2001). Some researchers noted a correlation between the growth of the pollen tube and the degradation of the nucellus. As the fertilized egg develops, it divides to form the embryo and the suspensor. The suspensor is a group of cells that hold the embryo and transport nutrients to it. The number of cells varies among species and sometimes even within a species. However, when the suspensor is no longer needed, it is eliminated by PCD. Also, sometimes more than one egg is fertilized (polyembryony); should that happen, all embryos except one are eliminated by PCD.

Pollen develops in the anther of the plant. PCD of the anthers allows the pollen to be released. The anther splits along a line in a process called dehiscence, in which the cells along the line die. Autophagy may be necessary for development of pollen, as research has shown that mutations that render autophagy defective in rice result in male sterility and immature pollen (Hanamata et al., 2020).

The germinating plant gains nutrients from the endosperm (part of the seed that contains starch and other nutrients) and the aleurone (protein granuals stored in the seed). In angiosperms, the endosperm, which contains mostly starch, is surrounded by the aleurone, which contains mostly protein (some of the pigments that color multicolor corn are in the aleurone as well). Once the endosperm is used up, the aleurone is subjected to PCD.

Xylem Formation

All vascular plants use a system of hollow tubes called xylem to move water from the roots to the rest of the plant. The tubes form when mesophyll cells called differentiating tracheary elements (TEs) form a second cell wall (Turner, Gallois, and Brown 2007). The phytohormones auxin and cytokinin and multiple genes have been implicated in the differentiation process. The TEs then undergo a form of PCD. The central vacuole collects various degradation enzymes, and then the tonoblast that surrounds the vacuole bursts to release those enzymes. The cell contents and part of the cell wall are destroyed. The resulting long tubes form the xylem network. The formation of xylem exhibits several characteristics of PCD. For example, genes that trigger cell destruction are upregulated. Cyclohexamide inhibits the process, indicating that protein synthesis is involved.

Root Cap

The growing parts of plants are called the meristem. The apical meristem is found in the growing tips of the stems and the roots. In the growing root tip, it is protected by a cap of cells. As the root tip pushes its way through the soil, the cells of the cap are continuously replenished as newer cells push past the older cells, and the older cells then die shortly thereafter. Damage from the abrasion of the soil is not the cause of the death of the cells, however. The cells die even if the root is growing in water. The cells are programmed to die, and they show classic features of PCD, including smaller cell bodies and fragmented DNA.

Trichomes

Trichomes are the small "hairs" found on the leaves and stems of most plants. They protect the plant against insect attacks and slow respiration. Programmed cell death is involved in the development of some trichomes and in the senescence of others, such as the leaves of tobacco (Uzelac et al. 2017).

Leaves

Leaves have many different shapes, and one might wonder if PCD is involved in producing those shapes. However, PCD turns out to be quite rare in leaf development (Gunawardena 2008). The two most often-cited examples, the Swiss cheese plant (*Monstera obliqua*) and the Madagascar lace plant (*Aponogeton madagascariensis*), use cell death to remodel their leaf blades, but the patterns produced are quite different. The Swiss cheese plant uses cell death to form gaps in the leaf that expand as the leaf grows. In the Madagascar lace plant, cell death yields many small holes that give the leaf its lacy appearance. Programmed cell death can also be seen in the formation of fistular leaves in the Welsh onion (*Allium fistulosum*). As the leaves develop, some of the tissue inside dies, leaving long empty tubes that facilitate the movement of gases throughout the plant (Ni et al. 2015).

Nevertheless, PCD is very important in leaves. Plants shed tissues once they have served their purpose: deciduous leaves turn brown and drop in the autumn, evergreens shed their leaves continuously, and ripe fruit drops to the ground. These processes are called abscission. Abscission also occurs to protect a plant: a damaged leaf may be released to conserve water. Infections by insects can induce abscission of affected leaves as well.

Abscission occurs in three steps. In the case of a deciduous tree's leaves falling in the autumn, the first step involves chlorophyll. As the amount of sunlight decreases in the fall, levels of chlorophyll also drop. In addition to being responsible for the green color, chlorophyll is a valuable material for plants. For example, it contains nitrogen that is needed for amino acids, nucleotides, and other key compounds. The plant degrades the chlorophyll to recycle these materials. Other pigments are slower to degrade, and the leaf color changes from green to yellow

Figure 10.1 Abscission. Abscission is a process in which a layer of cells at the base of a flower, leaf stem, or fruit dies in a genetically programmed manner to allow the object to detach. Abscission allowed the leaves and acorns in this photograph to fall from the tree.

Photograph by author.

or red. In the second step, an abscission or separation zone forms at the base of the petiole, and a protective layer is formed just below that. The cells then release suberin and lignin under the abscission zone to form a waterproof layer. Finally, the cells in the abscission zone import water, swell, and burst, causing the leaf to fall; in essence, the leaf is pushed off.

The trigger for abscission is not completely understood. Several candidates have been suggested. Reactive oxygen species might be involved. They are formed naturally during metabolism, but their levels increase during times of stress, such as cool temperatures or attacks by insects or other pathogens. Reactive oxygen species react with other cell components and may initiate the process that results in the leaves dropping. Finally, the hormones auxin and ethylene have been implicated in abscission. When levels of auxin decrease, plant cells are sensitive to ethylene; ethylene induces cellulase and polygalacturonase, which degrade plant cell walls.

Plant Pathogens and Defenses

Plants are susceptible to a myriad of pathogens. Various strains of bacteria, viruses, fungi, and other microorganisms use different modes of entry to infect plants. Viruses depend on insect or mechanical damage to get into plants. Viruses tend to be inside the cell, and bacteria are outside in the extracellular spaces (Van Doorn et al. 2011). Bacteria use the natural openings in plant leaves and other parts. They slip through the stomata (pores involved in gas exchange) or hydathodes (pores that allow water to escape from flowering plants) or through wounds to the plant. High levels of humidity favor bacterial infections in plants, and bacterial outbreaks in crop fields typically occur after rains. The bacteria use a secretion system to deliver "effector" proteins into a host cell (Xin et al. 2016). With these complex structures, bacteria inject the effector proteins into the cytoplasm by penetrating the inner and outer membranes of the plant cell. The proteins support disease by establishing an aqueous environment that promotes bacterial growth and suppresses the host cell defenses. Fungi use their haustoria (hairlike threads that are a little like roots) to penetrate plant cells and allow other pathogens to enter. In the Irish potato famine in the 1840s, more than a million people died from starvation. Potatoes were a major foodstuff for the Irish, but the potatoes were killed by a fungus-like microorganism called *Phytophthora infestans*. The organism attacks the plant's defenses by secreting hundreds of effectors. The effector proteins have different modes of action. For example, one called PexRD54 hinders autophagy (Dagdas et al. 2016).

Plants lack a complex adaptive immune system, such as that in mammals. However, they are hardly defenseless (Dangl and Jones 2001). They have several layers of defenses. First, a waxy cuticular layer provides a physical barrier against infectious agents, somewhat analogous to the way our skin protects us. Second, that cuticular layer contains antimicrobial compounds that can kill pathogens. Third, plants have a wound response that includes protease inhibitors and other compounds to discourage feeding on plant tissues. Fourth, plants release volatile compounds that attract beneficial insects that feed on the insects that eat plant tissues. Finally, plants have an effective innate immune system. They use a "zigzag" model of plant immunity (Jones and Dang 2006). In this model, plants use pattern recognition receptors to identify pathogens and nonpathogens by their molecular characteristics. The first component of this system involves disease resistance genes (R) that encode R proteins. The R proteins detect proteins associated with pathogens (e.g., effector proteins) and bind and degrade them. In turn, some pathogens release proteins that can interfere with the first phase of the immune response, and plants then respond with programmed cell death to limit the infection to a small area.

The interactions between plants and various infectious agents are governed by avirulence genes in the pathogens and the resistance genes in the plants. If either gene is missing or mutated, the result is disease. If both are present, the infection is usually short and elicits a hypersensitive response (HR). The HR causes necrotic lesions to form around the infected cells and so limits the infection.

Many proteins with similar functions in different organisms have similar amino acid sequences. These proteins are referred to as conserved. Plants have evolved other proteins called pattern-recognition receptors to identify the conserved markers. When they sense such an alien protein, plants activate an immune response called systemic acquired resistance. This process develops more slowly than HR, but unlike HR, it occurs throughout the entire plant. It is based on the accumulation of salicylic acid (a molecule related to aspirin) that acts as a signaling molecule to induce the activation of pathogenesis-related genes at the site of infection and in the other parts of the plant. Salicylate can also be converted to methyl salicylate, and that volatile molecule provides a mechanism to transmit the signal to other plants. Pathogen-related proteins have several modes of action. Some disrupt the cell walls of the bacteria or fungus that is attacking the plant. Others signal the infection to other cells. Still others attempt to physically isolate the infection by cross-linking molecules in the plant cell wall and laying down lignin, the compound that makes plant stems woody. (Interestingly, some of these pathogen-related proteins are also responsible for human allergies.)

This arms race between plants and plant pathogens has been going on for millennia. Plants have their defenses; pathogens continually evolve new ways to subvert those defenses; and in turn, plants evolve new defenses. As pathogens evolve to escape the plant immune system, the R proteins also evolve to attack the new pathogens. Pathogens secrete proteins and other small molecules called effectors that interfere with the plant's defenses. Plants recognize some effectors and launch their own immune responses. This "arms race" between pathogens and plants has been going on for as long as there have been plants and pathogens—probably hundreds of millions of years—and it will continue into the future. This is the so-called "Red Queen" metaphor. In Lewis Carroll's *Through the Looking Glass*, Alice has to keep running with the Red Queen just to stay in the same place. In like manner, plants evolve to avoid pathogens, and pathogens continue to develop new methods of infecting plants (Han 2019).

Programmed Cell Death in Plant Diseases

All three types of programmed cell death are deeply involved in the defenses that plants have against pathogens.

Apoptosis

Examples of apoptosis are relatively rare in plant disease. Sphingolipid and its derivatives initiate programmed cell death in response to disease (Dickman and Figueiredo 2013). In these cases, the HR demonstrates a number of the characteristics associated with apoptosis. Some forms of PCD in response to plant diseases show a process that seems to be a hybrid between apoptosis and necrosis (Khurana et al. 2005).

To attack plants, pathogenic fungi secrete phytotoxic compounds. Some are specific to a particular host, and others can attack a broad range of plants. In addition, some are signaling molecules. *Cochliobolus victoriae* secretes the toxin victorin, which attacks oats (Dickman and de Figueiredo 2013). This chlorinated cyclic pentapeptide activates plant defenses, PCD, and disease, inducing cell death or cell susceptibility. The fungus also attacks *Arabidopsis* and specifically an R protein called LOV1.

Among the differences between animals and plants is that plants lack caspases. However, they do have caspase-like activities in various enzymes. For example, the toxin from *Fusarium verticillioides,* fumonisin B1, causes PCD when it activates vacuolar processing enzyme (VPE). VPE is located in the plant vacuole, and the activation of this enzyme begins the breakdown of the vacuole.

Another fungal pathogen, *Fusarium oxysporum*, offers a good example of apoptosis in plant disease. It affects tomatoes, tobacco, legumes, cucurbits, sweet potatoes, and bananas. This fungus has many taxonomic subgroups, called formae speciales, to indicate that a particular form of the parasite is adapted to a specific host—for example, bananas are affected by *F. oxysporum* f. sp. *cubense.* The fungus infects the vascular system and affects the transport of water, causing wilting, browning of the vascular system, stunting, and damping-off (a term for the death of seeds or seedlings). In bananas, the resulting disease is a significant threat to fruit production worldwide.

In one study, animal genes that repress apoptosis were transferred into embryogenic cell suspensions of bananas (Paul et al. 2011). Plant lines were prepared from those cells. Those plants were then less susceptible than wild-type controls to infection by *F. oxysporum* f. sp. *cubense.* The wild-type plants showed apoptosis features by TUNEL assay, but the transformed plants did not.

Autophagy

Autophagy's role in plant pathology is complicated. Autophagy breaks down cell material into their basic building blocks so they can be recycled elsewhere in the plant, so it has a role both in plant cell death and in promoting plant cell survival. Autophagy can be induced in plants by a shortage of food and nutrients or by environmental stresses stresses, such as oxidation, high salt levels, and osmotic stress.

During autophagy, cell materials are sequestered in structures called autophagosomes and then delivered to acidic vacuoles, where the material is degraded. In some cases, receptor proteins and autophagy (e.g., ATG8) interact to load the damaged cell material into autophagosomes (Stolz, Ernst, and Dikic 2014).

When plants are infected by pathogens, the plants respond with their own defenses. The resistance gene *R* is induced (described earlier). The infected cells immediately activate an HR that kills the infected cells and inhibits virus replication (Komatsu et al. 2010). The HR comprises a slew of defenses, including the production of reactive oxygen species and the activation of other defensive genes. The strategy is to localize the infection and prevent it from proceeding to other cells in the plant.

When a plant is infected, autophagy has been proposed to initiate the HR PCD and vacuolar cell death and to inhibit necrosis in differentiated cells (Minina, Bozhkov, and Hofius 2014). For example, ATG6 and 7 are key genes for autophagy. Their loss enables infection of a close relative of the tobacco plant called *Nicotiana benthamiana* by tobacco mosaic virus (Lai et al. 2011).

Necrosis

PCD is one of the manifestations of HR, but killing of the cells is controlled by mechanisms other than those that inhibit viral replication. For example, when *Nicotiana benthamiana* was infected by the *Plantago asiatica* mosaic virus, PCD involved the activation of the genes for SGT1 and RAR1 and the MAPKKKα-MEK2 cascade (Komatsu et al. 2010). The two facets of HR in this system were modulated by different signals. All three proteins were required for necrosis, but the kinase cascade was not needed in the inhibition of viral replication. The results were the same when potato virus X was the pathogen.

Ferroptosis (described earlier in this chapter) is a type of necrosis assumed to be involved in stress-related situations. It does not seem to be involved in plant cell deaths during root development or reproduction. However, it is involved in deaths related to heat shock. Ferrostatins are small molecules that inhibit ferroptosis by blocking the actions of ROS. When *Arabidopsis thaliana* seedlings were exposed to ferrostatins, the root hairs were protected against temperatures that would normally cause them to die (Distéfano et al. 2017). The protection failed at even higher temperatures, which cause death by other mechanisms. Interestingly, caspase inhibitors also inhibit ferroptosis, but the substrates of those enzymes are not known. Although it has been conserved in plants and animals, the role of ferroptosis is still unclear.

11

Death in Bacteria, Fungi, and Protista

Even the simplest of eukaryotic life forms—the fungi, algae, and protozoa—are complex organisms. Their complexity is not merely that of eukaryotic cellular organization but is compounded further, as it is in higher plants and animals, by the fact that cells of eukaryotic organisms may contain viruses.

—Paul A. Lemke (1976)

Quorum Sensing by Bacteria

We rarely notice bacteria, but there are a lot of them. The numbers are absolutely staggering. Scientists estimate that there are 5×10^{30} bacteria on Earth (Whitman, Coleman, and Wieve 1998). That's a 5 followed by thirty zeros, or 5 nonillion. By contrast, there are *only* 7.5×10^9 (that's 7.5 billion) humans on Earth. These microscopic single-cell organisms are the most abundant living organisms on Earth (if viruses are not counted). The total carbon content in bacteria is nearly equal to that of plants. Half of the total mass of living organisms on Earth is microbial, and that estimate might even be low. In fact, if numbers were the only criterion, we would have to say that bacteria are the dominant life-form on Earth.

Bacteria are ubiquitous. They are found forty miles up in the atmosphere and at the bottom of the ocean. They exist on essentially every surface, in the soil, in water, and inside and on the surface of our bodies. Most are harmless. Many are useful. Some are critical for our digestion and health. Others are pathogenic and even deadly.

Bacteria are deceptively simple. They are prokaryotes, which means that they lack an organized nucleus and the other specialized cellular organelles that are found in eukaryotes, such as yeast, flies, plants, animals, and humans. Thus, prokaryotes, including bacteria, separated from eukaryotes very long ago in the history of evolution. In fact, mitochondria and chloroplasts, two organelles in eukaryotes that deal with energy, are thought to represent an ancient symbiotic relationship between formerly free-living prokaryotes and the eukaryotes. Mitochondria produce energy for all eukaryotes, and chloroplasts engage

The Biology of Death. Gary C. Howard, Oxford University Press. © Oxford University Press 2021.
DOI: 10.1093/oso/9780190687724.003.0011

in photosynthesis in plants. These organelles have DNA that is separate from that in the nucleus, and they also display characteristics that are more similar to bacteria.

It's hard to be a bacterium. They live in a hostile world. They run out of food or water. They face temperature extremes. They can die from mechanical forces or poisoning. They can become prey to other organisms. If they try to encroach on other strains of bacteria or even cells of their own strain, they may encounter bacteria-killing proteins. Their main predators are viruses called bacteriophages (or simply phages). The phages outnumber bacteria ten to one (Chibani-Chennoufi et al. 2004), and a single bacterium can be infected by as many as ten different phages. The typical outcome of an infection is that the virus usurps the cellular machinery of the bacterium and uses it to produce the components of a new crop of virus. Ultimately, the bacterial cell bursts and releases the new viruses to infect other bacteria. This results in an extraordinary amount of killing. As many as 15 percent of all of the bacteria in the oceans are killed every day by phage infections.

Of course, bacteria have not simply given up in the face of this massive killing. To survive, they have developed systems that help them to defend themselves. Like all other organisms, they are involved in an arms race with their enemy the phage, and have evolved systems to protect themselves against those infections. The bacteria have many mechanisms that interfere with the phage life cycle, including blocking the attachment of the phage to the bacterial surface; blocking injection of phage DNA; destroying phage DNA; restricting phage RNA transcription or phage development; and more.

In addition, the bacteria communicate with each other by exchanging diffusible signaling molecules. This process is called quorum sensing (QS) (Allocati et al. 2015). The signaling molecules include a compound with the ominous-sounding name extracellular death factor (EDF). In *E. coli*, EDF is a short peptide with just five amino acids (Asn-Asn-Trp-Asn-Asn-OH) (Kolodkin-Gal et al. 2007); *Pseudomonas aeruginosa* secretes the molecule 2-n-heptyl-4-hydroxyquinoline-N-oxide (HQNO). The EDF disrupts the cytochrome bc_1 complex, resulting in the production of large amounts of reactive oxygen species that cause cell lysis and DNA release (Hazan et al. 2016). Bacteria may use QS to sense multiple compounds and infer certain characteristics of their environment (Cornforth et al. 2014). The production of, uptake of, and response to these molecules are well understood.

Bacteria use QS to deal with many stressful situations. In the classic example, bacteria use QS to regulate the density of the colony. As the density passes a certain level, the colony throttles back on its growth and even allows some bacteria to die. In this way, even though some individuals are lost, the colony survives by living off the nutrients supplied from their dead comrades until conditions improve. Intriguingly, the bacteria seem to be communicating to regulate the density of the population by having some bacteria initiate a sequence that kills the cell.

As noted in previous chapters, programmed cell death (PCD) is a common phenomenon in multicellular organisms from fungi to plants to animals and humans. Cells are sacrificed to prevent the spread of infections or for developmental reasons. That's easy to comprehend in a multicellular organism, and it fits neatly into our understanding of natural selection. Genes that give an organism a survival advantage are more likely to mean that the organism survives and reproduces. However, bacteria are mostly single-cell organisms. Could they use some version of programmed cell death? How could a single-cell organism profit from PCD? The notion that one bacterium would sacrifice itself to save others seems counterintuitive. It would also go against what we know about natural selection. What evolutionary pressure would cause a free-living cell to sacrifice itself to ensure the survival of other free-living single cells? QS solves this problem by allowing the colony or population to behave like a single organism.

Amazingly, all free-living bacteria contain genes for cell suicide that are critical for their normal cell growth under different conditions. One system involves pairs of toxins and antidotes (Buts et al., 2005). The toxins include poison cellular processes or cleave mRNAs to hinder translation. The antidotes inhibit the toxin. They exist in a balance in the bacterial cell. When the toxin is needed, the proteolysis of the antidote occurs faster than its synthesis, and thus shifts the balance of the two. In this way, the bacterium can control stress to its system and can kill itself if necessary.

Bacteria often contain multiple toxin-antitoxin pairs, and they have many physiological effects. These systems can be divided into those that act through proteins and those that act through RNA antisense, an RNA molecule with the complementary sequence. The ToxIN Abi system from *Erwinia carotovora* subspecies *atroseptica* has a toxin-antitoxin pair (ToxN and ToxI, respectively) and involves a protein-RNA mechanism (Finerana et al. 2009). The ToxIN system is effective against a number of different phage types. As another example, the bacterium *Myxococcus xanthus* forms fruiting bodies for reproduction (Yamaguchi, Park, and Inouye 2011; Yamaguchi and Inouye 2016). The toxin MazF causes cell lysis of about 80 percent of the population, and fruiting bodies form in the remaining 20 percent. This sacrifice of some of the cells in the colony means that there will be enough nutrients for the remaining cells to survive and reproduce. If the gene for MazF is lost, the cells can no longer form fruiting bodies. The mazEF system is found in *E. coli* as well and is activated by starvation, heat shock, DNA damage, and other stresses (Ramisetty, Natarajan, and Sarojini 2013).

Bacteria use PCD for many purposes. For example, they use it to prevent infection of the population. An infected cell initiates a death program; the cell dies, but this prevents the phage that infected it from reproducing and infecting other bacteria. They also use PCD when environmental conditions are not conducive to growth. Under those circumstances, bacteria produce spores that can survive in a sort of suspended animation until conditions improve. Death is a part of that process as well. Sporulation begins with cell division or mitosis that yields two cells, one small and the other large. The small cell will eventually become the spore, and the large cell provides support for the small one. When the spore is mature, the large cell activates enzymes that degrade itself and release the spore.

Another example is provided by bacteria in the genus *Streptococcus*. They grow as a network of fine fillaments called hyphae in the soil. This vegetative structure is called a mycelium. When food is scarce, some of the bacteria undergo PCD, and the rest begin to form a fruiting body or mushroom that grows above ground. At this point, a second round of PCD results in the death of more fungal cells that provide extra nutrients for the developing system.

Bacteria are found everywhere. They are often thought of as swimming in aqueous environments, but most bacteria, except those in the ocean, settle on a surface and organize themselves into a biofilm (Persat et al. 2015). Biofilms are fascinating objects: that involve one or more types of organisms, such as bacteria, fungi, and protists, that form a collective. The organisms secrete polysaccharides that form a matrix, allowing the organisms to stick to each other, and often to the surface they are growing on; the matrix also encloses the bacteria and protects them from antibiotics and other insults. Biofilms grow on many types of surfaces in many different environments. Two familiar examples are dental plaque and pond scum. The common feature of biofilms is that they are wet. And they have been with us forever—there are fossils that seem to be biofilms that are 3.2 billion years old. PCD in bacteria is critical for the development of biofilms (Bayles 2007): as bacteria die, they release their DNA, and that DNA is used to produce the matrix that allows the bacteria to stick to each other and to the substrate.

In summary, bacteria seem to display processes with many similarities to PCD in eukaryotes. These genetic processes might even provide the evolutionary basis for PCD in eukaryotes (Bayles, 2014). Further research will resolve these issues.

Fungi

I've never trusted toadstools, but I suppose some must have their good point.
—The Cheshire Cat, in Lewis Carroll, *Alice in Wonderland*

The Cheshire Cat makes an interesting point. Mushrooms and many fungi are a bit mysterious. They seem to appear overnight. They grow on dead things and often cover living things with a layer of fuzz or slime. The good they do is hidden underground or in the molecules that make them up, but their grotesque appearance makes them beautiful and repulsive at the same time. This fascinating group of organisms contains more than 144,000 known species, and there are probably a million more that are as yet unknown. They include molds, yeasts, rusts, smuts, mildews, mushrooms, and toadstools. They can be single-celled, filamentous, or multicellular; some dimorphic species change from filamentous to single cells. They might be free-living, parasitic, or symbiotic with plants or animals. In all cases, they are heterotrophs: they cannot make their own food. They survive by living off organic matter. Some live very short lives, just a day or so, but fungal colonies can grow for a long time and become amazingly huge. In the Malheur National Forest in Oregon, a colony of *Armillaria ostoyae* covers 2,385 acres (nearly 4 square miles). Estimates of its age range from 2,000 to 8,500 years.

The fungi are critical to life on Earth. They help to break down dead organic matter that can be recycled into other living organisms. In doing so, they supply essential nutrients to most vascular plants through hyphae that surround and sometimes penetrate the plant cells to form mycorrhizae. They provide antibiotics and drugs (e.g., penicillin) and foods (e.g., mushrooms, beer, bread). They cause plant, animal, and human diseases. While relatively mild infections, such as ringworm, athlete's foot, and yeast, come easily to mind, some fungal diseases are deadly, especially to those who are immunocompromised. Nearly 1.5 million people die each year from infections by *Cryptococcus*, *Candida*, *Aspergillus*, and *Pneumocystis*.

Death in Fungi

Like animals and plants, fungi die. As individual organisms or even as colonies, they can be injured or run out of food and water. Fungi suffer from viral diseases. Mycoviruses are viruses that infect fungi (Son, Yu, and Kim 2015). The first mycovirus was found in *Agaricus bisporus*, the white mushrooms that are popular in salads and on pizza, and it results in badly formed mushrooms. Unlike the viruses of animals and plants, most mycoviruses do not have a way to infect cells from the outside. They are passed from one cell to another when cells join or when spores are formed. The infections often do not kill the host, and this observation has led scientists to suspect that the virus and fungus have evolved into some type of symbiotic relationship.

Yeast, like other fungi, do not have an immune system. However, they do have means of detecting and destroying unusual cellular RNA that might indicate that the cell's RNA is defective or that a virus (which has only RNA) has invaded the cell. In *Saccharomyces cerevisiae*, the enzyme Srn1p has evolved to be able to deal with new viral RNA as part of the evolutionary arms race that has been going on for eons (Rowley et al. 2016).

The relationship between bacteria and fungi is more complicated (Wargo and Hogan 2006). The two are often found together in the environment, and they have a great effect on each other. In some cases, the bacteria enable the fungi to produce fungal virulence determinants more efficiently. These determinants include the means to avoid host defenses, such as surface receptors that bind to host cells and outer coats that prevent the host from defending itself. In other cases, they inhibit the fungi. In yet other cases (such as biofilms), they work together in a system that has characteristics distinct from those of either species alone (de Boer 2017).

Programmed Cell Death in Fungi

Fungi rely on PCD in many facets of their life cycle, including development, responses to external threats, and recognizing the difference between self and non-self. The fungi also have the three major mechanisms of PCD: autophagy, apoptosis, and necrosis. During the development of multicellular fungi, cell death is used to eliminate unwanted cells so that further cell remodeling and differentiation can occur.

Many fungi are single cells rather than multicellular organisms. Thus, the death of a cell is, in effect, death of the whole organism. Still, some single-celled fungi undergo similar processes (Gonçalves et al. 2017). For example, *Saccharomyces cerevisiae* exhibits several of the classic features of apoptosis. ATPase is localized to the endoplasmic reticulum, phosphatidylserine migrates to the outer leaflet of the plasma membrane, and DNA becomes fragmented and condenses. *Schizosaccharomyces pombe* has other features typical of programmed cell death. Yet even single-celled fungi sometimes grow as colonies and exhibit a degree of differentiation. Dead zones sometimes form in the middle of a colony. If that dead zone is removed, growth slows at the periphery. This seems to indicate that the cell death in the dead zone somehow benefits the cells at the periphery.

The budding yeast *Saccharomyces cerevisiae* undergoes asymmetric cell division. A daughter cell buds off from a mother cell, leaving a bud scar on the mother cell. Mother cells die after 20–30 divisions, but the aging clock is re-set in the daughter cells.

PCD is involved in senescence and tissue remodeling (Lu 2006). In yeast, the expression of proapoptotic genes is increased by genetic defects, pheromones, oxygen stress, and aging. Mitochondria and the ubiquitin-proteasome system have roles. There are similarities to animals, including caspase-like and metacaspase genes and homologs to Bcl-2, AIF, BI-1, and IAP genes. Some species feature apoptosis after meiosis, the division of germ cells to produce sperm or eggs. Others do not.

Among the multicellular fungi, the filamentous fungi display characteristics of programmed cell death. For example, the ascomycete fungi *Coniochaete tetrasperma* and various *Neurospora* species use a cell death mechanism to reduce the number of spores per meiotic division from eight to four. A somewhat analogous process of nuclei reduction occurs in *Agaricus bisporus*. Some multicellular fungi show senescence as they age. In fact, the ascomycete *Podospora anserina* has been used as a model organism for aging studies for some time. As the fungus ages, mutations accumulate in the mitochondrial DNA, the activity of cytochrome oxidase c decreases and interestingly allows an extension of lifespan, the mycelium begins to slow its growth, pigmentation increases, and peripheral hyphae, the main vegetative structure of the fungus, die.

As filamentous fungi grow and move along surfaces, they encounter bacteria, fungal colonies of the same or different species, and other organisms. The ability to recognize fungi of the same species is important. When two filaments touch, the terminal cells fuse (anastomoses), allowing the cell contents to mix. If the two cells are not compatible, a typical cell death process results in the generation of hydrogen peroxide and the death of hyphae (Glass and Dementhon 2006). Having two nuclei of different species is called a heterokaryon ("hetero" for different and "karyon" for nucleus.) The process is controlled by genes called *het* (for heterokaryon incompatibility). The genes involved have been characterized molecularly. For example, in two ascomycetes, *Neurospora crassa* and *Podospora anserine*, the *het* genes have diverse sequences, do not belong to a single gene family, and have other cellular functions. In some cases, the *het* genes differ by only a single amino acid but are still incompatible. The significance of the heterokaryon incompatability is not known. Two hypotheses have been suggested: in one, the whole process is simply an evolutionary accident and has no real purpose, while in the other, this represents a non-self-recognition system that protects the fungus from mixing with an incompatible other fungus or from incorporating parasitic sequences from the other fungus. So far the evidence favors a non-self-recognition system (Saupe 2000).

Apoptosis

The proteins involved in apoptosis in fungi are similar to but simpler than those in animals (Sharon et al. 2009). For example, fewer proteins are involved, and the amino acid sequences of those proteins have little similarity (also called sequence homology) with the animal proteins. The function is similar, but the sequences are not.

Lu, Gallo, and Kues (2003) describe apoptosis during meiosis to generate spores in the ink-cap mushroom, *Coprinus cinereus*. The black spores are found in the gills on the underside of the cap of the mushroom. Typically, an amazing number of spores are produced by the nearly 10 million basidia (spore-producing cells) in the gills. However, sometimes something goes wrong in the process, and the spores fail to form. When the problem is detected early on, an apoptotic process is set in motion. The reason is not clear, but some speculate that the mushroom does not want to use valuable resources to follow up on flawed spores and so apoptosis is triggered. Even though the mushroom produces an enormous number of spores, the chance that one will actually grow is very low. Thus, the mushroom does not waste resources on imperfect spores.

Because mushrooms are fungi, most of the mushroom exists as threadlike hyphae that form the mycelium underground. The part that we see and sometimes eat is the fruiting body, where the spores form. The mycelium can live for a very long time; some are believed to be more than 2,000 years old. However, the fruiting body or mushroom exists only for a short time. The cell walls are made mostly of chitin and other polysaccharides. Chitin is used by many organisms. For example, crustaceans and insects use it to form their exoskeletons. As the fruiting bodies develop, the fungi produce enzymes called glycoside hydrolases that act on cell walls to allow the fruiting body to expand and grow. Once the spores have been released, the fruiting body undergoes autolysis, and enzymes degrade the cell walls. Zhou et al. (2015) found that three of these glycoside hydrolases and chitinases may act synergistically to break down the cell walls in *C. cinerea*.

The maturation and autolysis of mushrooms involve numerous changes. Levels of volatile compounds rise and fall. Levels of enzymes that control the expansion of the stipe, the stalk-like structure under the cap and then the autolysis also change. And underlying all of these changes are dramatic changes in the expression of genes that control the enzymes. For example, in *C. cinerea*, a gene called *exp1*, which likely encodes an HMG1/2-like protein (these proteins are often involved in transcription control or chromatin remodeling), is strongly induced in the final three hours of development (Muraguchi et al. 2008).

After mushrooms are harvested, they are still alive, but the autolysis program continues and results in the browning of the gills and a loss of freshness in the mushroom (Sakamoto et al. 2012). The mushroom also begins to soften as the cell walls are broken down by various chitinases and other hydrolases. In like manner, a number of genes involved with the normal functioning of the fungus, development of the fruiting body, and spore development are downregulated.

Senescence can be observed in some fungal strains after they have been held in cell culture for multiple generations (Maheshwari and Navaraj 2007). For example, *Neurospora* strains normally grow as multinucleate mycelium. To reproduce, the nuclei begin to separate, and they form spores that germinate to form new fungi. However, as the fungi age, they develop intramitochondrial linear or circular plasmids (small sequences of DNA that are separate from the genes on the chromosomes) that integrate into the mitochondrial DNA, rendering the mitochondria dysfunctional. Interestingly, the dysfunctional mitochondria actually replicate faster than the functional ones, and so they spread through the multinucleate mycelium. Spores with dysfunctional mitochondria may have trouble producing energy from food (called oxidative phosphorylation). As a result, they "starve" to death. In any case, the culture fails to thrive and eventually dies out.

Philipp et al. (2013) screened *Podospora anserine* for genes that are activated with aging. They found that the expression of genes linked to ribosomes and the proteasome system was decreased with aging. This would make it difficult for the organism to make the proteins that it needs and to dispose of worn out proteins. However, certain other genes were enhanced, including those connected to autophagy. They concluded that autophagy may compensate as a quality control system when activity of the proteasome system is reduced by aging.

Autophagy

Autophagy is used by some pathogenic fungi to facilitate infection of the host. The filamentous fungus *Magnaporthe grisea* causes rice blast disease. The strategy used by the fungus involves brute strength. It begins when fungal spores are disseminated and germinate to form a germ tube and then undergo mitosis (splitting in two). One daughter nucleus migrates to the site of the original spore. The other moves to the other end of the germ tube, where a special structure called the appressorium is forming. The dome-shaped appressorium swells with water until it physically breaks the cell wall of the rice plant. Once that is done, a mycelium threads its way into the cell, and the infection begins. Although the mechanism is not yet clear, autophagy causes spore collapse and is required for a successful infection (Veneault-Fourrey et al. 2006; Veneault-Fourrey and Talbot 2007). That does not seem to be the case in *Podospora anserina*. In that species, autophagy is not necessary.

Protista

If you put a drop of pond water under a microscope, you might see several protozoans that are familiar to anyone who took high school biology, including amoebas, paramecia, single-cell algae, euglena, and volvox. Protozoans (members of the kingdom Protista) are a very large and loose group—they are not animals, plants, or fungi, and the origins and relationships of these organisms are far from well understood. Some members of this diverse group of single-cell organisms or colonies of single-cell organisms live in the water; others live on land. Some photosynthesize. Some are free-living. Others are parasites, and among them are several responsible for human diseases, including malaria, sleeping sickness, amoebic dysentery, and trichomoniasis.

The protista diverged from the other major groups a very long time ago, evolutionarily speaking. Nevertheless, a type of PCD has been documented in several protozoans, including *Plasmodium*, *Trypanosoma*, *Leishmania*, and *Giardia*, all serious human parasites (Kaczanowski, Sajid, and Reece 2011). Genomic analysis shows that the protista have genes that are typically associated with PCD. They respond to stresses by displaying several apoptotic-like features. For example, three apoptosis-related proteins (e.g., AIF, Endo G1, and cytochrome c) move from the mitochondria to the cytoplasm. The actions of those proteins are similar to the actions of similar proteins in animals. Unlike animals, protozoans lack caspases. Like plants, they contain metacaspases that enhance proteolysis. However, the exact function of the metacaspases is unknown.

Since protozoans are primarily single cells, one might ask about the value, if any, of PCD to these groups. Evolution is based on natural selection. Cells and organisms that are more fit reproduce more statistically. How can an individual cell benefit by dying for the mass? Is this really a form of PCD, an evolutionary accident, or a consequence of cellular aging? The question is particularly difficult for the free-living protozoans. Parasitic protozoans might benefit from the ability to control the number of parasites living on a host. If the host survived longer, more protozoan offspring could ultimately be produced. The argument assumes that all of the parasites are from the same clone. It would not work as well in situations with infections by multiple clones.

Some protista are not always single cells. For example, *Dictyostelium discoideum* exists both as single cells and as multicellular organisms under different circumstances (Nelson and Baehrecka 2014). This free-living amoeba feeds on bacteria living in the soil and in moist leaf litter. As long as there are plenty of bacteria, the *Dictyostelium* continues to eat and reproduce by mitosis. If the bacteria are depleted, the amoebas begin to aggregate into a motile pseudoplasmodium of about 100,000 cells that is referred to as a "slug." The slug creeps along over the substrate looking for a more favorable environment. Once such a place is found, the slug begins to differentiate further into a fruiting body. In this process, about 20 percent of the cells in the slug form the stalk that supports the fruiting body. Two steps are involved. First, the cells begin the process of autophagy to preserve resources. Homologs for the autophagy signal mTOR are found in *Dictyostelium*, but how they are involved in the first step is not clear. Second, a differentiation-inducing factor causes the cells to die. Both signals are needed to induce cell death. The fruiting body then produces spores that continue the growth of the colony.

Other lines of evidence show the involvement of apoptosis in protista. For example, *Leishmania amazonensis* use the compound phosphatidylserine to avoid the immune system of the host so they can infect host cells (Wanderley et al. 2013). The infected cells contain a lot of apoptotic parasites that then die to allow healthy parasites to survive. The metacaspase in *Toxoplasma gondii* is critical to apoptosis. Genetic knockout of the gene for the enzyme resulted in less apoptosis, and overexpression of the enzyme increased in more apoptosis (Li et al. 2016).

While these examples are interesting, not everyone shares the belief that PCD is involved in protista biology. While they agree that there are some similarities in protista and animals, they note that some key factors are missing (Proto, Coombs, and Mottram 2013). They conclude that the cell death in most parasitic protista can be attributed to either necrosis or just coincidence. The features of PCD in protista are not exactly the same as those in animals, and in most cases, it is counterintuitive to imagine PCD in single-cell organisms. Nevertheless, the evidence for PCD is significant.

12

Death on a Grand Scale

The history of life is a story of massive removal followed by differentiation within a few surviving stocks, not the conventional tale of
steadily increasing excellence, complexity, and diversity.
—Stephen Jay Gould, *Wonderful Life: The Burgess
Shale and the Nature of History*

Dinosaurs dominated the Earth for about 130 million years. About 66 million
years ago, they became extinct. From an evolutionary standpoint, their passing
was sudden. They were not alone—nearly 75 percent of all plant and animal
species on Earth became extinct in the same mass extinction event (Renne
et al. 2013).

Death on that scale is incomprehensible. We live on a planet that is teeming
with life that exhibits an amazing degree of diversity. Plankton, kelp, fish, and
blue whales fill the seas. Sulfur bacteria and a vast collection of worms inhabit
the salt flats at the sea's edge. Dandelions, maidenhair ferns, lichens, towering
redwoods, and fields of wheat cover the land. Iguanas, African elephants, earthworms, and slime molds live on the land. Dragonflies, hummingbirds, and eagles
fill the skies. Even we humans are covered with an astounding array of bacteria,
nematodes, and other microscopic fellow travelers.

Yet, as remarkable as it is, this diversity represents only a fraction of all of the
species that have ever lived on Earth. Life began on Earth at least 3.5 billion years
ago, and extinction is a common occurrence. Extinctions occur continuously;
for whatever reason, a species cannot adapt quickly enough to the changes in
the environment surrounding it, and it dies out. Today we rightly fear the loss
of particular species, such as mountain gorillas, blue whales, and Siberian tigers. However, species have been lost almost continually since life began. In fact,
90 percent of the species that have ever existed have gone extinct. And while it
might sound tragic, the extinction of some groups of species changed the course
of evolution and allowed other groups to blossom. Stephen Jay Gould wrote,
"Alter any early event, ever so slightly and without apparent importance at the
time, and evolution cascades into a radically different channel" (1989, 51).

The Biology of Death. Gary C. Howard, Oxford University Press. © Oxford University Press 2021.
DOI: 10.1093/oso/9780190687724.003.0012

Table 12.1 Major Mass Extinction Events

Event	When (millions of years ago)
Cambrian-Ordovician	524
Ordovician-Silurian	447–443
Late Devonian	375–359
Permian	251
Triassic-Jurassic	201
Cretaceous-Paleogene	66
Current?	Ongoing

Mass extinctions are something completely different. In these events, more than 75 percent of all species are lost over a short period ("short," that is, in geological or evolutionary time), usually about 1 million years. This is many orders of magnitude beyond our experience. None of us has seen anything like it, and hopefully we never will. However, mass extinctions have happened several times in the history of the Earth (see Table 12.1), and these events have resulted in important transitions in the history of living organisms. For example, the mass extinction that ended the reign of the dinosaurs and let the mammals expand, something that eventually led to the emergence of humans. In most cases, the causes of the mass extinctions are not precisely known. Volcanos, climate change, asteroid strikes, and other causes have been invoked. Unfortunately, we might be in another one right now, and that one is directly due to the activities of humans.

Geology and Biology

To understand mass extinctions, it is helpful to understand how biology and geology are related. Throughout the time that life was evolving, the Earth itself was also changing. The surface of the Earth is covered with seven or eight major plates and a number of minor ones; plates consist of a relatively light continental crust and a somewhat denser oceanic crust that moves over the surface of the earth. This movement, called continental drift, is extremely slow, but over millions of years the plates can move long distances. The enormous North American plate includes all of North America, part of Siberia, and Greenland. The motion of the plates is powered by spreading in the middle of the oceans as new material wells up from the Earth's mantle; where plates meet, subduction occurs (the edge of one plate moves under the edge of the other). Along those edges, earthquakes and volcanoes are common. Ultimately, the energy for the movements comes from the Earth's cooling but still fantastically hot mantle and core.

The continents have not always existed in their current form, and they will not look the same far in the future. There is a rhythm to it all: over hundreds of millions of years, a single supercontinent breaks apart into multiple smaller continents, the continents drift apart, and eventually they come together again to form another single supercontinent.

Scientists believe that at least three supercontinents have existed over the past 2 billion years. The oldest known so far, Nuna, was about 1.8 billion years ago. Rodinia was about 1 billion years ago. Pangaea was about 300 million years ago. In between those three, varying numbers of smaller continents, such as we have today, existed. Some of the evidence for this movement come from rocks containing iron. Iron can be magnetized, and when the rock was molten, the magnetized iron oriented itself along the lines of the Earth's magnetic field at that time. Once the rock solidified, its history was locked in stone and can be compared to the magnetic orientation of other rocks.

Most computer models suggest that in a few hundred million years, Asia and North and South America will collide to form a new supercontinent. In one (Mitchell, Kilian, and Evans 2012), the Caribbean Sea and the Arctic Ocean will disappear, and South America will wind up alongside the eastern coast of North America.

Continental drift is not the direct cause of extinctions, but it is often a significant factor. In addition, the geology provides a timeline for the expansions and extinctions of life that have occurred over millions of years. What kinds of species existed at specific times? In the very beginning, there was no life at all on Earth. The continents were completely barren. Eventually life formed in the oceans, but the land was still empty. Sometime after that, life moved onto land as well. As the continents broke apart, animal and plant populations became isolated from one another, and new species developed.

In the remainder of this chapter, we will describe the mass extinctions that have occurred in the context of their geological time and the evolution of species at that time.

Defining Mass Extinctions

Defining mass extinctions is not quite as simple as it might seem. In general terms, mass extinctions are geologically brief (occurring over a period of about 1 million years) and result in the loss of many species in a broad range of habitats on land and in the seas around the world. When graphed as number of lost species versus time, they represent sharp peaks above the "normal" level of extinctions that are occurring continually. Thus, defining a mass extinction event can be a bit arbitrary. However, scientists have generally agreed on five main events.

Examinations of the fossil record are critical for establishing the timelines needed to sort out these events. These include fossils of all types. For example, Song et al. (2011) examined 10,000 microfossils from seven regions of south China. Stable oxygen isotope data from the mineral parts of marine animal fossils are also very helpful; when alive, the marine animals incorporate oxygen isotopes into the calcite, aragonite, and apatite that form their shells and other hard parts (Schneebeili-Herman 2012). The ratio of those isotopes can be used to determine the water temperature, and those data can then be associated with other information to estimate a date.

Causes of Mass Extinctions

No one really knows what causes mass extinctions. It's one of the most intriguing questions in biology, and more than one theory has been suggested to explain it. Whatever caused the extinctions, it had to be global and extraordinary—no local or small-scale event could account for the wiping out of 75 percent or more of all species existing on the Earth at a given time. In addition, species in both the oceans and on land were affected. Clearly, radical changes were involved. Possible factors might be large-scale changes to climate, acidification of the oceans, significant changes in ocean temperature, or changes in the depth of the shallow regions of the oceans near the continents.

Even if the exact cause is unclear, we can speculate on what happened. For example, climate change and global warming are often assumed to be involved. For most plants, temperatures consistently above 40°C (104°F) are lethal. For animals, the upper limit is 45°C (113°F). At those temperatures, the proteins that make up the organism are damaged. Many organisms produce substances, called heat shock proteins, that help to protect the organism's other proteins from extreme conditions. These heat-shock proteins help proteins that are damaged or denatured to regain their normal conformation, which is critical to their function. Ocean animals suffer at lower temperatures (above 35°C), for different reasons: as the temperature rises, their need for oxygen goes up, but the amount of oxygen in the water goes down. Those sea creatures that are very active and need lots of oxygen (e.g., cephalopods, such as octopus and squid) are the most vulnerable.

High temperatures have other effects too. For example, they can increase the activity of organisms that are involved in decomposing dead animals and plants (fungi and bacteria), which in turn can increase the amount of carbon in the atmosphere and reduce the fertility of soils. That seemed to have happened to the Early Triassic soils of Australia and Antarctica (the reduction in the productivity of plants around this time is indicated by the loss of peat swamps during this period).

Not all species were equally affected by these factors. Longrich (2012) examined the loss of lizards and snakes at the Cretaceous–Paleogene boundary (about 66 million years ago), around the time of the extinction of the dinosaurs and many other species. He found that nearly 83 percent of the species in these two large groupings were lost. Those that survived tended to be smaller and have a large range.

The most famous theory about what triggered this particular mass extinction concerns a meteor that struck the Earth. This theory was suggested in 1980 by Luis and Walter Alvarez, a father and son team at the University of California, Berkeley (Alvarez et al. 1980). The impact is thought to have sent countless tons of dust and debris into the atmosphere, blocking out the sun for years. The resulting endless "winter"-like climate would have cooled the earth for many years. And indeed, a meteor impact crater was found on the northern edge of the Yucatan Peninsula in present-day Mexico. Named for the nearest town, the Chicxulub crater is about 110 miles in diameter and 12 miles deep. The meteor itself was estimated to have been about 6 miles in diameter. The crater is shaped like an egg, suggesting that the asteroid hit at a shallow angle (i.e., 20°–30°) toward the northwest. Additional evidence comes from a layer of rock strata rich in the element iridium that was deposited around the world at exactly that time; such high concentrations of iridium are thought to result from a meteor impact. There is other geological evidence, too, including the presence of shocked quartz (formed when quartz crystals experience a sudden rise in pressure) and tektites (formed when the energy of a meteor strike melts rock that reforms as droplets) near the impact site. The strike has been dated to a bit over 66 million years ago, and that date coincides fairly well with the beginning of the extinction event.

While this model has gained wide public acceptance and is taught in most public schools, many scientists have questioned it from the beginning. Their criticisms fall into several categories. For example, to cause such an extinction event, a strike has to be large enough and hit at a time preceding the event by the appropriate amount. The impact at Chicxulub was very large and seems to have occurred at more or less the right time. There have certainly been enough other impacts large enough (Napier 2014), but none have the appropriate timing to account for the other extinction events. Furthermore, other massive strikes did not always result in such an event. For example, neither the strike that created Chesapeake Bay nor another in Quebec left any impression in the fossil record.

The evidence for the impact setting off this particular extinction is very strong. However, is there anything else that could account for the widespread disruption of ecosystems needed to explain mass extinction events? The primary alternative theory involves very large volcanic eruptions. To be large enough, these eruptions would have to be far larger than any seen in recorded history.

Indeed, some very large eruptions have occurred. One began right before the beginning of this extinction. It formed what is called the Deccan Traps in west-central India (Schoene et al. 2015). The lava covered as much as 800,000 square miles, and in some areas is as much as a mile thick. Aerosols and ash could have easily been ejected high enough to make it into the stratosphere, causing a cooling of the planet over very large areas and even perhaps worldwide. There have been other very large eruptions from volcanos (see Figure 12.1) over the life of the Earth; the eruptions that created the Siberian Traps and the Central Atlantic Magmatic Province both deposited vast amounts of basalt, and their timing might associate them with other extinction events (Napier 2015). Unfortunately, the standard methods for fixing dates do not offer enough precision to establish a solid correlation between these eruptions and extinction events (Erwin 2014).

Figure 12.1 Volcanic activity. The eruption of volcanoes has been suggested as a cause of the mass extinctions that Earth has experienced. This photograph shows lava flows from the Pu'u 'Ō'ō vent of Kilauea on the island of Hawaii.
Photograph by author.

Major Mass Extinction Events

Cambrian-Ordovician Extinction

During the geologic period called the Cambrian Explosion, the size, diversity, and complexity of living organisms on the Earth expanded rapidly (at least in evolutionary terms). During this period the basic forms of most of the current phyla evolved. Beginning about 524 million years ago, however, a series of mass extinction events (the end-Botomian event, 524 to 517 million years ago; the Dresbachian event, 502 million years ago; and the Cambrian-Ordovician event, 488 million years ago) ended the Cambrian Explosion and eliminated the bulk of the species in existence at that time.

Stephen Jay Gould wrote eloquently about this period in his book, *Wonderful Life: The Burgess Shale and the Nature of History*. The Burgess shale deposit in Canada holds an unusual collection of the fossilized remains of soft-bodied animals from that time, and evidence found there suggests that there were phyla existing at the time that no longer are present on the Earth. As Gould wrote: "Some fifteen to twenty Burgess species cannot be allied with any known group, and should probably be classified as separate phyla. Magnify some of them beyond the few centimeters of their actual size, and you are on the set of a science-fiction film." Other paleontologists have taken some issue with his assessment, but his enthusiasm for the diversity of life is inspiring.

Ordovician-Silurian Mass Extinction

Until the Silurian (410 to 444 million years ago), life existed mostly in the seas. Most animals in the Ordovician (444 to 485 million years ago) were still invertebrates. These included the common trilobites; bryozoans, also called "moss animals" because of their moss-like appearance; brachiopods, which look like clams (but are not related to them) and are found today in very cold waters; and graptolites, thought to be members of the Hemichordata, early relatives of animals with a backbone. The first vertebrates, fish, had appeared by then, but they lacked jaws and bones; those would evolve later, in the Silurian. The Silurian would also feature a major step in evolution when multicellular organisms began to move from the sea onto the land.

The Ordovician-Silurian mass extinction occurred about 447 to 443 million years ago and had two major times of dying, but they were within a few hundred thousand years of each other. The cause of the extinction is not completely understood, but we do know there were global environmental changes at the very end of the Ordovician, perhaps associated with the presence of the supercontinent

Gondwanaland near the South Pole (Finnegan et al. 2012). The first of the two pulses of extinction seems to be associated with glaciation (Sheehan 2001). Sea levels fell in the areas near the continents, resulting in harsh climates in the low and middle latitudes, and deep ocean currents moved nutrients and toxins to shallower areas. The animals that survived had to adapt to new conditions. When the glaciers eventually receded in the early Silurian, sea levels rose, the climate moderated, and oceanic circulation stagnated; the result was another pulse of extinction.

Late Devonian Mass Extinction

The Late Devonian mass extinction event occurred 375–359 million years ago. Some scientists attribute it to two events, but others believe that it was a series of smaller events over that span. Regardless of how it happened, the result is clear: 79–87 percent of all species disappeared. However, the effects were not felt equally across the board. Some groups fared better than others. Armored fish died out completely. Many marine species were badly affected, particularly in the shallow tropical seas. Coral reefs and their inhabitants were particularly hard hit. Yet plants and insects fared pretty well.

As with most of the mass extinctions, the cause of the Late Devonian event is not clear. The environment was undergoing extensive changes at the time; levels of oxygen dropped dramatically, sea levels rose, and the earth cooled, triggering a new ice age. It's possible that the plants that survived the event so well might have had a role in triggering it. Roots had just evolved, and their action might have contributed to the weathering of rocks and the building of soil. This allowed nutrients to be washed into the ocean, and resulted in algal blooms that depleted the oxygen in the seas. The plants might also have captured increasing amounts of carbon dioxide, lowering global temperatures overall.

Of course, the two usual suspects might have had a role—there were very large volcanic eruptions in what is now Vilnuy province in eastern Siberia, and multiple meteorite impacts. Unfortunately, none of those fits well chronologically with the extinctions.

Permian Mass Extinction

The Permian mass extinction was the most lethal of the last 542 million years (Burgess, Bowring, and Shen 2014). It took place about 251 million years ago, after the Carboniferous Period, when the presence of huge swamps allowed many new species to evolve, including amphibians and shortly thereafter reptiles. Sharks filled the oceans.

In this mass extinction, nearly 96 percent of all species were killed off. Marine species were particularly impacted. The remaining trilobites disappeared. Many corals were lost along with the predatory eurypterids found in shallow waters. Land animals also suffered. Notably, this was the only mass extinction to affect insects. Land plants largely escaped, except for forests. All of the living organisms on Earth today, including humans, are descended from the few species that survived the Permian mass extinction.

Several factors were probably involved. At the time, the landmass was concentrated into a single supercontinent called Pangaea. It covered the Earth from the North Pole to the South Pole, and conditions in the interior were very hot and dry. Sun et al. (2012) reported that these exceptionally high temperatures—the highest ever estimated to have existed on Earth—had a significant effect on life in the regions near the equator. Based on work by Schneebeili-Hermann (2012), who noted that the amounts of oxygen isotopes in marine organisms can be used as a proxy for sea-surface temperatures (when these creatures make their shells, they incorporate oxygen into the exoskeletons and other hard parts), Joachimski et al. (2012) showed that global sea-surface temperatures in the late Permian might have reached more than 32°C, conditions that would put many species under stress.

On top of this, a familiar additional stress appeared: a huge volcanic eruption in Siberia, with lava flows that covered about 2 million square kilometers. Enormous amounts of gases were released, including huge quantities of carbon dioxide, and this caused more heating. Methane frozen in marine layers might have melted, adding another greenhouse gas to the mix.

Triassic-Jurassic Mass Extinction

The Triassic-Jurassic mass extinction occurred about 201 million years ago. The period before the event was very productive: vertebrates thrived, new species of sharks and fish appeared, marine reptiles became significant predators in both the shallow waters and the seas, and dinosaurs developed (first as small dog-sized animals, but by the end of the epoch many had become giants).

Toward the end of the Triassic, about 18 million years ago, two or three extinctions seem to have combined to cause the Triassic-Jurassic mass extinction event. About 76–84 percent of all species were lost. These included mammal-like reptiles, large amphibians, and many dinosaurs. Marine life was not spared: the corals and reef species were hit hard again, conodonts, ammonites, and brachiopods were affected, and over 90 percent of bivalve species were lost.

The causes of this extinction are the least well understood of the major events. Climate change, eruptions, and an asteroid impact have all been blamed for this loss of life. We do know that a rift was opening up between Europe and North America, with its accompanying volcanic action, and the Atlantic Ocean was beginning to form. Sea levels fell globally, and loss of the warm shallow seas hurt reefs. There was less water in the oceans and less surface water on land, and the oxygen levels in the water were lower. A series of large volcanic eruptions occurred over a period of about 500,000 years toward the end of the Triassic, creating what's known as the Central Atlantic Magmatic Province.

Blackburn et al. (2013) used uranium-lead testing (which determines ratios of isotopes of these two elements in various rock samples) to refine the timing of the age and duration of volcanism in this period. Their work showed a series of four peaks in the amount of volcanism that coincide with the mass extinctions during that time. The most significant result of the extinction event was that the great loss of both land and marine organisms enabled the dinosaurs to become the dominant species on Earth for the next 130 million years.

Cretaceous-Paleogene Mass Extinction

The Jurassic was a great time for evolution. The mass extinction at the end of the Triassic had eliminated most of the large amphibians and mammal-like reptiles. In that environment, the dinosaurs flourished on the land and even moved into the sea and air, developing into hundreds of different species. Flowering plants and birds were first seen.

Then about 66 million years ago, the Cretaceous-Tertiary mass extinction hit. Most famously, the dinosaurs disappeared. But many other animal groups, plants, and plankton did too. In fact, about 71–81 percent of all species died. Most larger organisms perished, but among the ones that survived in greater numbers were insects, small mammals, birds, flowering plants, fish, corals, and mollusks.

As we've seen, it's commonly thought that a very large asteroid struck the earth, starting a chain of events that led to this mass extinction. Renne et al. (2013) examined samples of the tektites from around the Chicxulub crater, and they also tested bentonites, which are found in volcanic ash, to determine the timing of volcanic eruptions from about the same time. They then compared these data with changes in the levels of carbon in the atmosphere and the numbers of mammals in the western Williston Basin, which is in eastern Montana and western North and South Dakota. They found that the Chicxulub impact occurred before the extinctions, but that ecosystems were already critically stressed; their conclusion is that the impact and its sequelae likely pushed many species over the edge.

The environmental stresses that existed before the impact came from an Earth undergoing significant transformation. The continents looked more and more like they do today, but sea levels were falling rapidly. Volcanoes were very active, producing lava, carbon dioxide and other gases, and ash. The world began to warm. Estimates are that the average temperature might have increased by 10°C and the temperature of the oceans by 4°C. The oceans would have become more acidic. Phytoplankton died, and then so did the larger marine animals that eat them. On land, plants, herbivores, and predators suffered and died. Then the asteroid hit.

Mass Extinctions Now and in the Future

It's clear that mass extinctions have killed very large numbers of species multiple times throughout the history of Earth. Those die-offs have changed the course of life on Earth. Species that had evolved to dominate the land or the seas and had done so successfully for millions of years were wiped out in geologically short order. The rules of the game changed so fast that they could not survive. Death ended their dominance. But in all cases—at least so far—life came back to thrive with new species as the dominant forms.

What lessons can we learn from those events? One is that we might be causing a modern-day mass extinction. Another is that we might be sealing our own fate as part of the bargain.

In recent years, many large animals have been brought to the verge of extinction. Others are on the endangered species list. Conservationists have gone to great lengths to protect Siberian tigers, gorillas, blue whales, African elephants, and rhinoceros. But it's already too late for the many species that have been lost since humans arrived on the scene. Wooly mammoths and passenger pigeons are well-known cases, but there are many more that are not so famous. In some cases, humans have hunted the animals to death. In others, loss of habitat is the cause.

The direct cause of most of the mass extinctions in the past is still unknown, though the impacts of space objects, huge volcanic eruptions, climate change, acidification, and changes in the levels of the seas have all been invoked as possible factors, either alone or in combination. But one difference between those events and the possible sixth mass extinction that lies before us is painfully clear: we humans are responsible for the changes that are happening all around us right now.

Anthony Barnosky, curator of the museum of paleontology at the University of California, Berkeley, and his group (Barnosky et al. 2011) reviewed the loss of species over the last few millennia and compared the fossil record and the modern data on extinctions. They found that extinction rates from today are higher than expected, based on the fossil record. For example, before humans expanded so much in the last 500 years, mammals went extinct at a rate of only two species per million years. However, in the last 500 years, at least 80 of 5,570 species of mammals have gone extinct—a much higher rate. They conclude that the Earth is heading for a sixth mass extinction within the next 300 to 2,000 years.

The most shocking speculation about the possibility of a mass extinction in the relatively near future is whether we might be setting ourselves up for extinction along with the other life on Earth. Global warming has become a critical issue in the world today. Climate scientists are convinced that this is an extremely serious problem, one that requires immediate attention. Many scientists think the oceans have absorbed the worst effects of climate change so far (Bijma et al. 2013; Payne and Clapham 2012). The excess carbon dioxide in the atmosphere has been dissolved in the ocean. As a result, the pH of the oceans has fallen, oxygen levels have dropped, and the composition, structure, and functions of marine ecosystems have been affected. These factors are all symptoms of changes in the carbon cycle that we know occurred in past extinction events. These conditions are further exacerbated by human effects, such as pollution, eutrophication (excessive growth that limits oxygen in the water), and overfishing. So far we have seen lower numbers of living organisms in the oceans and especially a decline in sensitive species.

Humans have taken advantage of the carbon fuels laid down millions of years ago to power our modern world. Coal, oil, and gas have been burned in increasing amounts over the last couple of centuries. As a result, greenhouse gases (primarily carbon dioxide) have accumulated in the atmosphere, and global temperatures are on the rise. Nearly all climate scientists attribute these increases to human activity. The Intergovernmental Panel on Climate Change (IPCC) estimates that the Earth's mean surface temperature could rise 2°–6°C by the end of the twenty-first century. While that may not sound threatening to an average person, in fact it is a very serious rise for most biological systems.

Even now weather patterns are changing in ways that are predicted by models of climate change. Weather events are more extreme, with fiercer storms. The glaciers and ice sheets at the poles are melting. Plants and animals will have to try to adapt to these new conditions, but the real concern is that, at some point, we will pass a threshold at which the feedback systems of global warming will become unstoppable. Then the temperatures on Earth will start a runaway rise. At that point, instead of the mass extinction changing the course of evolution, we will have ended evolution by killing all life on Earth forever.

13

Last Hominid Standing

Run the tape again, and let the tiny twig of *Homo sapiens* expire in
Africa. Other hominids may have stood on the threshold of what
we know as human possibilities. . . . One little twig on the mammalian branch, a lineage with interesting possibilities that were never
realized, joins the vast majority of species in extinction. So what?
Most possibilities are never realized, and who will ever know the
difference?

—Stephen Jay Gould, *Wonderful Life*

Life on Earth has survived at least five major extinction events. In each case,
much of the life on Earth was lost. Our earliest ancestors survived all of these
events back to the beginning of life. Since our hominid ancestors evolved
a few million years ago, they have not had to face a mass extinction event,
although there have been some near misses (as with the Toba eruption,
described later). Still, it has not been an easy time for hominids or any other
organism. In nature, competition is fierce, and natural selection is relentless.
Like untold numbers of species before them, many early hominid branches
of our family tree disappeared. Either they went extinct or—more likely—
were subsumed into modern humans by interbreeding. As Gould said, we
will never be able to "rerun the tape" to discover what potential they might
have had. We, *Homo sapiens*, are the winners, at least for now, and even we
had several close calls.

The traditional view is that those other species simply went extinct, and we are
the last hominid standing. A more interesting explanation is that most of those
other species were actually part of the same family, and interbreeding resulted in
the combination species that we are today. Understanding human evolution is
complicated by what anthropologist Henry Gilbert (2018) refers to as the "species problem." Scientists have identified multiple species in the genus *Homo*, to
which modern humans belong, and in the genera of the predecessors of *Homo*
(such as *Australopithecus*) In most cases, we assume that one species is defined as

The Biology of Death. Gary C. Howard, Oxford University Press. © Oxford University Press 2021.
DOI: 10.1093/oso/9780190687724.003.0013

reproductively separated from others—that it cannot interbreed with a member of another species. However, this is not the case. First, the International Code for Zoological Nomenclature has no standard for defining a species. A group is accepted as a species if it is anatomically distinct from other groups. Second, the fossil record is incomplete and often difficult (perhaps even impossible) to interpret in terms of reproductive ability. Thus, multiple species of *Homo* that existed as modern humans began to appear and migrate should perhaps be thought of as markers of a location or time rather than as distinct species. In fact, it is quite likely that all of what we now consider to be species in the genus *Homo* were actually a single species with some locale-based variations. In this chapter, we will use the term "species" in that sense to track the movement of the various branches of modern humans.

What Happened to Our Predecessors?

Our earliest ancestors appeared as much as 20 million years ago, and the hominid line has existed for around 8 million years. Even 2 million years ago, there were at least fifteen different species of hominids.

Homo sapiens likely evolved from *H. erectus* across the Mid- to Late Pleistocene in Africa. As noted by Gilbert (2018), there is considerable controversy as to whether the various groupings of hominids before that point were separate species, and if so, whether they should be placed in the genus *Homo* or in the genus *Australopithecus*. For example, *H. habilis* lived between 2.4 and 1.4 million years ago. They had long arms and a face that was more ape-like. Some doubt that these should be classified in *Homo*. As another example, *Homo floresiensis* lived in Indonesia from 95,000 to 17,000 years ago. They were small-statured and had a very small brain. What happened to them? Did they die out? Were they simply one more step in the evolution of modern humans? Or were they always part of the same group that became modern humans?

Only *H. sapiens* remains today. Our "cousins" either died out or were bred out of the lineage. Yet even *H. sapiens* itself was nearly lost before it left Africa and might have been reduced to fewer than 10,000 individuals. It's conceivable that, had there been another twist of fate, it could easily have been another of those earlier species that survived to become "modern humans," and our *H. sapiens* line might only exist as some DNA sequences. We will look more closely at three groups: Neandertals, Denisovans, and *Homo erectus*. We know a little more about the Neandertals than other groups, but we really know little about what caused them all to go extinct.

Our Possible Family Members

Neandertals

Our close relatives the Neandertals (*Homo neanderthalensis*) existed alongside us (*Homo sapiens*) for some time. We even bred with them, so the genomes of many of today's humans have some amount of Neandertal DNA (typically 2–4 percent). They left Africa before *H. sapiens*. What happened to the Neandertals? Mostly it's unknown.

For more than 200,000 years, the Neandertals lived as the dominant hominid species in Europe, the Middle East, and Western and Central Siberia. In some areas, their diet was heavy with meat, such as wooly rhinoceros and wild sheep. In other areas, such as Gibraltar, they ate mostly mushrooms, pine cones, and moss. Weyrich et al. (2017) made these determinations by examining the plaque on teeth of Neandertals in those areas. In some cases, they found that the Neandertals likely self-medicated by chewing poplar, which contains salicylic acid (related to aspirin), and moldy material, containing *Penicillium* (the mold from which the antibiotic penicillin is derived). Then the Neandertals disappeared.

The Pleistocene (2.5 million to 11,770 years ago) was characterized by great temperature changes that put tremendous pressure on many species of plants and animals, and that might have been a source of pressure for Neandertals too. In Europe in particular, where the mountains tend to run east to west, organisms (including Neandertals) had less opportunity to expand or contract north to south when seeking more favored temperatures, as the mountains blocked migration routes. Neandertals survived in isolated pockets, such as the Iberian peninsula (Hofreiter and Stewart 2009).

About 42,000 years ago, one of the last groups of Neandertals lost their battle for survival in the caves of Gibraltar. Gibraltar was different then. Because of the last Ice Age, much of the Earth's water was frozen at the poles and in glaciers over much of the Northern Hemisphere. Thus the ocean level was 250 feet lower than it is today, and a long fertile plain stretched out below what are today oceanside cliffs. The climate was cooler, but it was still relatively mild at Gibraltar, and plants and animals thrived there, providing the Neandertals with a steady supply of food. At Gibraltar, the Neandertals lived in at least ten sites.

At about the same time the Neandertals at Gibraltar died out, the few other groups of Neandertals remaining in Europe and Eurasia also died out. Around that time, though, a new group of hominids arrived in Europe, and for a few thousand years *Homo sapiens* lived with the Neandertals in the same ranges. At the end of that time, *H. sapiens* was well established in Europe and the Neandertals were lost. What happened to our close relatives?

The reasons are the subject of considerable discussion among scientists, and several ideas have been suggested. First, relatively short periods of radical climate changes in Europe might have threatened food supplies. At that time, the Northern Hemisphere experienced several Dansgaard–Oeschger cold cycles (Staubwasser et al. 2018), which were cold cycles of a few decades separated by longer warm cycles.

Second, some researchers have assumed that the two groups actually fought each other over hunting rights and that over time the less successful Neandertals simply lost the wars. Neandertal skeletons show a large number of injuries (Zollikofer et al. 2002; Jiménez-Brobeil and Oumaoui 2009), though they might be partially attributable to the rigors of a hunter-gatherer lifestyle that involved taking large prey at close quarters. Sala et al. (2015) described another example of two certainly fatal wounds in one skeleton. The two wounds might suggest conflict between groups.

Third, some have suggested that while modern humans developed a division of labor between men and women, Neandertals did not do so, and this might have been a factor in the demise of the Neandertals. Kuhn and Stiner (2006) pointed out that men and women had similar responsibilities in early hunter-gatherer societies, but in the Upper Paleolithic period, humans began to specialize according to gender. Men specialized in hunting, and women began to focus on gathering, cooking, and childcare.

Fourth, the extinction of the Neandertals may involve dogs. Dogs and humans have been partners for thousands of years. Interestingly, Pat Shipman (2014) credits this relationship as a possible reason that early humans were able to outcompete their Neandertal neighbors. She suggests that about 40,000 years ago, humans, Neandertals, and wolves all were predators on large herbivores, such as mammoths, rhinoceros, and bison. The loss of those mammals coincides with the arrival of modern humans. The development of complex projectile weapons (the use of which requires complex skill [Zhua and Binghamb 2011]) and the domestication of dogs may have given modern humans an advantage over Neandertals and other older groups. Shipman believes the alliance between wolves and humans began when humans entered Europe. The wolf-dogs would have protected the humans from other carnivores that might have tried to steal their supply of meat. They would also have chased and harassed prey. Once the prey was exhausted, the humans could more easily kill them with spears or maybe bows and arrows.

Her theory of this alliance is not without challenges. First, even Shipman admits that this pushes the date for domestication of dogs back much further than previous suggestions. Most scientists believe that dogs were domesticated about the time humans began to use agriculture, about 10,000 years ago. Others, including Perri, Smith, and Bosch (2015), are intrigued but not convinced by

Shipman's arguments. To them, the evidence is not compelling regarding the nature of the mammoth bone sites. They also point out that wolves typically attack the youngest and oldest large prey to conserve energy and lessen the chance of injury while taking down large prey, and they are not likely to have attacked such a massive animal as a mammoth.

Another theory to explain the extinction of the Neandertals was recently advanced by Vaesen et al. (2019), who note that Neandertal populations were small even before the arrival of modern humans and suggest that Neandertals died out because of a combination of small population size, inbreeding, and fluctuations in birth and death rates and sex ratios.

Denisovans

Much less is known about the Denisovans (*Homo denisova*) than about the Neandertals. Only a few Denisovan bones were collected from a cave in south-central Siberia in 2010. However, their genome has been determined. They and the Neandertals are thought to have branched off from the group that eventually developed into *H. sapiens*. Denisovans split off from Neandertals about 400,000 to 450,000 years ago (Posth et al. 2017). Denisovan nuclear DNA is more closely related to Neandertals than humans (Reich et al. 2010), but humans and Neandertals have more similar mitochondrial DNA. The exact relationships of these groups are not known. We do believe that interbreeding was not uncommon—*H. sapiens* with Neanderthals and Denisovans, Neanderthals with Denisovans, and Denisovans with a mystery hominid species. Unfortunately, the fate of the Denisovans is even less well understood than that of the Neandertals. But they are important because of their relationship to modern humans.

Homo erectus

H. erectus is a fascinating and extremely successful group. They existed from 1.9 million to 140,000 years ago. Their bodies indicate that they had completely transformed from living in the trees to living on the ground. They were upright and walked on two legs. They had a larger brain than earlier hominids, and they made stone tools and used fire. *H. erectus* migrated out of Africa long before *H. sapiens* did and spread across Europe and Asia to Spain, China, Indonesia, and South Africa.

Their relationship to other hominid species is not clear; they might be an ancestor to several other species, including *H. heidelbergensis*, *H. antecessor*, *H. neanderthalensis*, *H. denisova*, and *H. sapiens*. These are all described as separate species on the basis of morphology. However, most overlapped with the others in time, and all of them interbred, making the puzzle of their family tree even more difficult to unscramble.

So what happened to *H. erectus*? It lasted a long time and was quite successful. Did it die out or not? A recent paper (Shipton et al. 2018) seemed to imply that, based on evidence from one site, *H. erectus* was conservative in its willingness to adopt new technologies, which perhaps led to its extinction. But that characterization does not seem to be consistent with a species that invented hand tools and fire and that spread from Africa to Central and South Asia. We know that *H. erectus* only colonized areas below 55° latitude and that *H. sapiens* spread beyond that line into more challenging environments. That might indicate that *H. erectus* was less capable of adapting to certain types of changes, and that changing environments might have put additional stress on them.

A couple of theoretical considerations might shed light on the fate of species. One possibility is that *H. erectus* might be a chronospecies with some or all of the other hominid species. Chronospecies occur when development involves a continual and uniform change from an ancestral form (*H. erectus*) to a later form (*H. sapiens*). The separation between them can be hard to detect, especially when the fossil record is sparse. In addition, the transition from *H. erectus* to *H. sapiens* might involve punctuated rather than continuous evolution. That is, *H. erectus* might have undergone few or no evolutionary changes for a long period of time (evolutionary stasis). Then the transition to *H. sapiens* might have been rapid in evolutionary terms.

Homo heidelbergensis

H. heidelbergensis appeared around 780,000 years ago (Buck and Stringer 2014). Their bodies were stockier than those of other early hominids, and this body shape allowed them to deal with colder temperatures. Thus, they could migrate into areas that others could not. They were the first hominid to build permanent shelters. They are thought to be the first hominid to lack air sacs, which might suggest that they had a greater ability to use language than previous groups. Air sacs are large cavities located just above the vocal cords in all of the apes, except humans. Their function is unknown, but their lack has been hypothesized to be associated with the development of language. *H. heidelbergensis* was also predominantly right-handed, a trait associated with greater language capabilities. *H. heidelbergensis* is also a possible candidate for the last common ancestor for *H. sapiens* and the Neandertals. They died out about 200,000 years ago.

Emergence of *Homo sapiens* and Interbreeding
with Other Hominids

The appearance of *H. sapiens* is complicated and unclear. W. W. Howells (1976) put it best: "That part of human history covering the emergence of modern man and his regional differentiation continues to be surprisingly obscure. Locations of some elements of agreement or controversy . . . have long been clear, but the dimensions of the whole problem are far from obvious. The trees are familiar, but the forest is not."

Most scientists believe that modern humans appeared about 200,000 years ago. While some humans and other hominids remained in Africa, others left in several waves that were mostly unsuccessful in that they pretty much died out (Rabett 2018; Nielsen et al. 2017); remains found in Israel are likely some of these unsuccessful migrations (Hershkovitz et al. 2018). The short-statured *H. floresiensis*, remains of which were found on what is now called Flores Island in Indonesia, coexisted with modern humans in that area. Interestingly, Flores Island is now home to a modern group of pygmy humans. Tucci et al. (2018) examined the genomes of ten modern individuals and genetic variations among thirty-two individuals and concluded that modern humans on Flores Island carry DNA sequences from Neandertals and Denisovans and that short stature has evolved independently at least twice there.

The Neandertals are probably our best-known hominid relatives. We know that they bred with modern humans because all non-African humans today have some percentage of Neandertal. The average is about 2–4 percent (Sankararaman et al. 2016), but one individual was reported to have 6–9 percent Neandertal DNA (Fu et al. 2015). In fact, many of us also carry other admixtures of genes of extinct groups, such as Denisovans.

That these groups could interbreed is not surprising. Our hominid family tree is complicated and not completely worked out, but one theory posits that we were originally one species called *H. heidelbergensis*. According to this theory, about 500,000–600,000 years ago, some left Africa for Europe and Eurasia and became Neandertals. Others migrated into East Asia and became Denisovans. Those left in Africa developed into *H. sapiens*. Then about 70,000 years ago, some of the *H. sapiens* left Africa, and these were able to mate with the Neandertals and Denisovans.

In addition to the interbreeding that went on outside Africa, interbreeding was also taking place in Africa between modern humans and earlier groups (Hammer et al. 2011). Though studies were hindered by the difficulty of examining fossil DNA from Africa (the tropical climate causes DNA to deteriorate more quickly there), a computational and statistical approach helped them pinpoint three regions in the genomes of modern African populations (totaling about 2–3 percent of the total sequence) that are likely archaic DNA. This clearly shows that interbreeding was common in Africa before modern humans left there for Europe and Asia.

To make the picture even more complicated, humans in Africa interbred with another ancient hominid group both before and after the ancestors of European and Asian populations split off and migrated away. By the scientists' estimates, DNA from that unknown group now makes up somewhere between 4 percent and 8 percent of modern human ancestry (Ragsdale and Gravel 2019).

As we've seen, the relationship between modern humans and our close relatives the Neandertals is complicated. Interestingly, though, while the nuclear DNA of Denisovans and Neandertals are more similar to each other than to the DNA of modern humans, the mitochondrial DNAs of modern humans and Neandertals are more similar. How can that be explained? One piece of evidence comes from a femur found in southwestern Germany. The bone was that of a Neandertal, but it seems to be from a different group of Neandertals than those usually examined. It contains different DNA sequences that might indicate that a small group of humans left Africa before the main exodus. By its mitochondrial DNA, it split from the others 270,000 years ago (Posth et al. 2017). An alternative explanation is that around 250,000 years ago, a group of modern humans may have left Africa. They interbred with the Neandertals, who had left Africa 200,000 years before that. That group of modern humans was small and left no DNA sequences in the Neandertal nuclear DNA. However, they did leave some sequences in the mitochondrial DNA. Of course, there are other possible explanations.

Other recent evidence also points to multiple migrations out of Africa and considerable interbreeding. Fossils found at a site called Jebel Irhoud in Morocco were dated to 300,000 years ago and suggest that modern humans may have been around for about 100,000 years longer than previously thought. Scientists from the Max Planck Institute for Evolutionary Anthropology and the Moroccan National Institute for Archaeology and Heritage concluded that before modern humans left Africa for the rest of the world, they spread throughout the African continent (Hublin et al. 2017; Richter et al. 2017).

Leaving Africa

Around 60,000–70,000 years ago, humans left Africa and began to spread around the world. The most likely route out of Africa was across the Bab-el-Mandeb Strait of the Red Sea. Today it is twelve miles across water from Yemen to Djibouti. Back then, though, sea level was more than 200 feet lower than now, and the strait would have been much narrower, with maybe some small islands in between, making crossing much easier. Another possible route was from northern Egypt to the Sinai.

The reasons humans left Africa are not clear, but climate may be a factor. About that time, the Earth's climate changed dramatically; it was one of the coldest periods of the last Ice Age. Some scientists believe that during this time the human population dropped to around 10,000. As temperatures warmed, humans expanded in numbers and territory. By about 50,000 years ago, they had made it to Southeast Asia and Australia. Others went into Europe and on to the Americas.

The region referred to as Island Southeast Asia and New Guinea is one of the most complex for human evolutionary history (Tucci et al. 2018). Modern Papuans carry DNA sequences from two distinct Denisovan groups that separated 350,000 years ago (Jacobs et al. 2019). A third Denisovan lineage occurs in modern East Asians. Clearly, modern humans were interbreeding with Denisovans as much as they were with Neandertals.

Genetic Bottlenecks and Human Evolution

Our own *H. sapiens* family was nearly wiped out on several occasions. The causes are typically not known, but the fact that the populations came close to disappearing can be detected by an examination of genetic diversity, which allows scientists to reconstruct effective population size.

Genetic diversity refers to the variability in the genetic information among individuals in a particular population. The degree of diversity is related to the ability of a population to adapt to changes in the environment. The more diversity in the group, the greater is their ability to survive. Diversity is partially related to the number of individuals, but not entirely.

One part of the puzzle is that humans have surprisingly little diversity. Our DNA sequences are very closely matched, and we should have more. For example, the human line separated from the chimpanzee line 5 million to 6 million years ago, and since then chimps have developed a very large degree of genetic diversity—far more than modern humans have (Gibbons 1995).

Humans began in Africa, and the greatest amount of human genetic diversity exists there, among the people who remained there. That is the reason that the consensus of scientists is that modern humans emerged in Africa. The people who migrated from Africa to populate the rest of the world have much less diversity, even though they had some genetic additions from the Neandertal and Denisovan populations.

What happened to us? Some scientists believe that human populations suffered some severe setbacks over the millennia during which the population was greatly reduced. Those close calls are referred to as "bottlenecks." A bottleneck results from some event (such as climate change, a large-scale volcanic eruption, or an outbreak of a deadly disease) that reduces the population significantly. Along with the loss of population can come a loss of genetic diversity—something that diminishes the population's ability to withstand further dramatic change. (For those wanting a more detailed explanation of the relationship between genetic diversity and population size, see Amos and Harwood 1998.)

To study genetic diversity in humans, scientists look for subtle differences between groups. In particular, two types of sequences have been used to estimate genetic diversity. First, nearly half of our DNA is made up of short sequences that sometimes move around. These "transposons" jump from one region to another, and one of the most common groups is the Alu sequences—there are over a million copies in our genome. Second, microsatellites are short sequences of a few nucleotides that are repeated many times. These are found throughout the human genome, and their mutation rate is higher than for the rest of our DNA.

The concept of a genetic bottleneck has supporters and detractors. There are four main events that have been suggested as possible bottlenecks for humans.

Bottleneck Candidate 1: 1–2 Million Years Ago

One group of scientists believes that humans experienced one of these bottleneck events about 1 million years ago (Huff et al. 2010), at the very beginning of *H. sapiens*. Using Alu sequences, they estimated the effective population size of archaic *H. sapiens* and its closely related groups *H. erectus* and *H. ergaster* was about 18,500 (with a total population of about 55,000). In contrast, our effective population size today is about 1 million (with a total population of more than 7 billion). They conclude that our species suffered a catastrophic event about 1 million years ago that nearly wiped us out. Hawks et al. (2000) report a genetic bottleneck in the *H. sapiens* lineage at about 2 million years ago but see no evidence of a more recent bottleneck.

Bottleneck Candidate 2: 170,000–195,000 Years Ago

Another possibility for a major bottleneck might have occurred between 170,000 and 195,000 years ago. The Earth was getting colder at that time, as another glacial stage called Marine Isotope Stage 6 began. Glaciers increased in size, and the loss of water to the glaciers also increased the size of deserts. Whatever happened, it was very hard on the human population. Humans were reduced from an effective population size of 10,000 to perhaps only 600. With so few humans, each group's location was critical. One of those was at Pinnacle Point Cave 13B (Mossel Bay, Western Cape Province, South Africa). This location provided a wealth of nutritious root crops and protein-rich shellfish (Marean et al. 2010), minerals for tools, and a moderate climate.

Bottleneck Candidate 3: 70,000 Years Ago

The third bottleneck candidate occurred more recently, about 70,000 years ago, at the time when some modern humans were starting to leave Africa. The first evidence of this was that, as we have seen, as a species humans have much less genetic diversity than other species, far less than we might expect. Zhivotovsky, Rosenberg, and Feldman (2003) examined 377 microsatellite markers from fifty-two regions in the genome and found a very low level of genetic diversity that indicates some kind of disaster within the last 100,000 years. One possible candidate is the eruption of the Toba volcano on the island of Sumatra in Indonesia about 75,000 years ago.

When normal volcanos erupt, they affect only the local area. In a large explosion, tons of rock and debris can be injected into the atmosphere and drift downwind; still, the affected area is relatively small. However, Toba was a supervolcano. It was the largest known eruption ever, and threw massive quantities of rock dust into the atmosphere. This dimmed the sun for six years, and may have cooled the Earth for 1,000 years. Winters grew longer. Plants died and then so did the animals that depended on them. The effective human population size might have been reduced to as little as 1,000 (Robock et al. 2009). However, the coastal environment of Southern Africa seems to have been a sanctuary from whatever challenges were initiated by the Toba eruption (Smith et al. 2018).

Not everyone agrees with this explanation. They point out that there are stone tools above and below the layer of dust in neighboring India after the eruption, and so humans clearly survived in some numbers. The fossil record seems to show that animals also survived. Yost et al. (2018) claim that there was no long-term cooling or effects from the Toba eruption and that no bottleneck occurred.

Bottleneck Candidate 4: 5,000–7,000 Years Ago

Karmin et al. (2015) suggested that a bottleneck involving the Y chromosome might have occurred 5,000–7,000 years ago. The effective population of males was reduced to about 5 percent of its original level. However, mitochondrial

DNA indicates that the numbers of females continued to increase. What could have caused this result? It's not likely to have been ecological or climatic factors; these might affect birth ratios, but those differences are usually too small to account for the mismatch. And the populations were too large to implicate founder effects. Zeng, Aw, and Feldman (2018) showed that culture and sociology are possibilities. For example, competition between patrilineal kin groups that broke out into warfare may have resulted in a loss of males, and consequently a loss of Y chromosome lines.

Last of the Hominids

As we've seen, more than 99 percent of all species that ever existed are extinct. Extinctions have been commonplace throughout history. Many extinctions were relatively gradual, the result of environmental shifts, evolutionary "arms races," or local competition between species. Others were relatively sudden, caused by asteroid strikes, large-scale volcano eruptions, or other causes. What are the chances that it could happen again, with humans being among the extinct species this time? Using the history of modern humans, Snyder-Beattie, Ord, and Bonsall (2019) calculated the odds that humanity would go extinct from natural causes as less than 1 in 14,000 and likely less than 1 in 87,000; if they expand that calculation to use earlier hominid ancestors, the odds drop even further, to just 1 in 870,000. However, this estimate takes into account only natural events. Human-caused events, such as nuclear war or the climate crisis, could reduce those odds considerably.

We are the last of the hominids. Modern humans haven't yet been around as long as some of the others were, such as the Neandertals and *H. erectus*. It could have turned out differently; our lineage, with its interesting possibilities, might have joined the vast majority of species in extinction. But we got lucky. We had some abilities that other groups may have lacked, including better tools and likely a better command of speech and language. We spread further and into more environments than any other hominid species. Yet even so, we were lucky. Several times we came close to joining the vast majority of species in extinction.

While we have some tantalizing clues, we still don't have definite answers to many of the most intriguing questions: What is our family tree? How are we related to the other groups, especially those with whom we seem to have interbred? Why did the other groups disappear? Did they really disappear or are they still with us in our genetic makeup? What is so special about us? How long will we last? Might there be a successor to us? Of course, we will not be around to hear the answers to those last two questions.

14

Bioethics

Death is terrible, but still more terrible is the feeling that you might
live forever and never die.

—Anton Chekhov, *Note-Book* (1921)

Who Wants to Live Forever?

Who wants to live forever? Amazingly, not everyone does, even if good health
is assured (Freund, Nikitin, and Ritter 2009). When 252 adults aged eighteen to
seventy-seven years were asked in a nonrepresentative survey if they would take
a pill that would let them live forever, 77 percent said no. When they were told
the pill would ensure them twenty-five additional healthy years, only 47 percent
said they would take it. The answers of why were surprising. Among those who
said no, some cited religious objections. Others still believed old age would be
filled with limitations and pains. Still others thought they could complete everything they needed to do in a normal lifetime. Those who said yes wanted to
see what would happen in the future or wanted extra time to do more. Others
simply enjoyed life. For questions like those, context is clearly important. Old age
involves not just physical factors but a social construct of aging (Vincent 2006).
People might think of old age in terms of what they know, which is typically that
the human body begins to fail with age. The survey did not take into account the
current personal situation of the respondents, though this variable might partially explain the results: happy people are probably more likely to want to live
longer, while those with health, financial, or other challenges might be less en-
thusiastic about facing more of the same in the future.

Humans change over their lifetimes. An extension of life would likely be more
attractive to older people than younger ones (Freund, Nikitin, and Ritter 2009).
Young people are focused on developing, learning, gaining the experience necessary for their future endeavors, and deciding on life goals. In the normal course
of events, older persons have completed those steps and their life's work. They
gradually let go of that work and their personal goals and focus more on leisure. If life span is to be extended, significant changes will be needed in personal
goals, the expectation and availability of work, and social interactions. These are

The Biology of Death. Gary C. Howard, Oxford University Press. © Oxford University Press 2021.
DOI: 10.1093/oso/9780190687724.003.0014

significant changes for society, but they are not insurmountable. In fact, life expectancy has increased rather dramatically in the last 100 years, and society and individuals have managed to adjust.

Of course, people are not completely consistent. This particular survey showed lukewarm support for extending life, but consumer demand suggests a different story. The demand for products that slow the effects of aging is huge and growing. People spend amazing amounts on products to wash away the gray or to remove wrinkles. Requests for cosmetic surgery are at an all-time high. So consumers are voting with their dollars, and they want to at least look younger, even if they can't stay that way.

Partridge and Hall (2007) differentiate between two types of research that have implications for human life span. In one kind of research, the goal is not the direct extension of life span, but rather developing or improving treatments of aging-related diseases, or preventing those diseases in the first place. If successful, such research would have the effect of extending the life span and health span of individuals. Even that brings some questions. For example, what is the appropriate use of resources to treat older and sicker patients? Is it right to attempt to save every patient, regardless of cost? In our current system, do these extreme measures siphon off scarce resources from where they could be better used?

The second type of research focuses on directly extending life span by studying the causes of aging. It too raises questions. Should we be interfering with nature or God's will? Should we continue to expand human life span to more than, say, 120 years? What are the social consequences?

Treating Current Patients

In a speech in 1984, Richard D. Lamm, then the governor of Colorado, set off a firestorm about healthcare and the elderly. Although the governor disputed accounts of his comments, his words were interpreted by the media as meaning the old and ill had a duty to the next generation to simply die and get out of the way (*New York Times*, March 29, 1984). The passionate and in many cases overwrought response was not helpful, but the issue is a serious one.

Patients near death raise a myriad of ethical issues that concern the practical delivery of care. Many of the sickest patients wind up in intensive care units in hospitals, and some die. Ethical dilemmas surrounding end-of-life care are a common occurrence (Curtis and Vincent 2010) and affect physicians, nurses, families, and others. Often the patient is too sick to be an effective participant in the discussion. The stress from these decisions adds to the burnout among the staff, in particular. They are often the ones who must make the ethical distinction between continuing critical care and withdrawing it. Families can be completely overwhelmed by the situation. They might agree not to begin treatment, but they have difficulty taking it away once it has begun.

In the last decades, choice has become much more of a factor in medical care. Various rights movements have inspired patients to assert their desires for treatment, or to not be treated. Patients are participating in discussions about their options to a greater extent. This is particularly true for older patients. In addition, medical science has made great advances in the care of older patients. Longer life spans create the desire for even longer and healthier lives, and the cycle continues.

Still, there are situations in which the patient's choice is limited by decisions made by the physician (Kaufman, Shim, and Russ 2006). First, outside factors place limits on what is available to patients. Third-party payers will pay for only certain procedures and drugs. And in some cases physicians may be threatened by litigation based on their choices about treatment. Second, the standard of practice for some diseases effectively limits options. Third, patients and families are sometimes reluctant to take the responsibility for making care-related decisions, particularly life-and-death ones for older patients. They prefer to leave those decisions to the physician.

Not all of the ethical dilemmas involve end-of-life decisions about the elderly. Some of the most challenging and heartbreaking occur in the neonatal intensive care unit, where the decisions concern whether to continue therapies for newborns with little or no chance of survival. In a study of 122 neonatal intensive care units in Europe, nearly every physician interviewed had participated in these discussions and decisions. The stress of these decisions on the staff, especially the nursing staff, is enormous.

Birth defects are one reason parents and medical personnel may need to make decisions about whether to institute or continue care for newborns. Another problem is often seen with premature infants, whose immature lungs lack sufficient surfactant to work effectively and who may suffer from respiratory distress. These infants may wind up with too little oxygen and too much carbon dioxide in the blood (hypoxemia and hypercarbia, respectively) and a blood pH that is too low (acidosis). The development of artificial surfactants was a major breakthrough in this area. A third common problem with premature babies is sepsis, which occurs in about 25 percent of babies weighing less than 1,500 grams and is life-threatening. This systemic inflammatory response is caused by infections of *Staphylococcus*, *Streptococcus*, or gram-negative bacteria. Premature infants have poorly developed immune systems, and the frequent placement of needed lines into blood vessels and blood draws provide opportunities for infections to begin. Necrotizing enterocolitis is a fourth potential problem in premature babies. It occurs when the barrier provided by the gastrointestinal mucosa is compromised. A breach in the mucosa can allow intestinal contents to seep into the abdominal

214 THE BIOLOGY OF DEATH

cavity and result in massive infection. Treatment often involves removal of the affected tissue and can result in long-term nutritional problems for infants who survive. And brain injuries are another hazard for babies born too early: bleeding in the brain (intraventricular hemorrhage) can destroy brain cells, and cerebral palsy can result in motor and cognitive deficits.

Long-term complications can result not just from these problems themselves but also from treatments for those conditions. For example, nearly 30 percent of infants needing supplemental oxygen develop bronchopulmonary dysplasia; in later life, they often have recurring lung infections and poor growth. Oxygen therapy may also result in retinopathy of prematurity, increased risk of cataracts, impaired vision, and various other eye disorders. Finally, as they grow through childhood, premature infants may have lower IQ scores, developmental delays, learning disabilities, poor social skills, or attention-deficit/hyperactivity disorder.

Ethical Questions

A number of bioethical questions arise from the efforts to defer death and extend life (McConnel and Turner 2005). Some of these questions are similar to those involved in developing any medical intervention. First, there are practical questions about new therapies. How can therapies that extend life be tested? Clearly, testing would take a long time. Who will assume the risks? These are good questions, but they also often apply to current disease research. Most people would not want that research to stop, and so why should aging research stop?

Second, there is a question of fairness. At least initially, the cost of interventions to extend life span would likely be considerable and so only available to the wealthy. While it must be recognized that medical care is not universally available today, allowing the wealthy to live longer would present serious ethical questions. Should some people be denied the chance to live longer or more healthily just because all cannot?

Resource scarcity is another aspect of fairness (Wareham 2015; Harbut 2019). As global populations increase, so will the competition for resources, such as food and clean water. Already populations are on the move because of the climate crisis. In 2018, the World Bank estimated that Latin America, sub-Saharan Africa, and Southeast Asia will generate 143 million more migrants by 2050 (Rigaud et al. 2018, 2). Dealing with these changes will require significant changes to government policies and personal expectations around quality of life.

For much of human existence, people had little choice in when they died. Today new technologies and the medicalization of death allow patients a degree of choice (Hetzler and Dugdale 2018). Future technologies to extend life will likely be very expensive, further exacerbating the divide between the haves and have-nots. Is that an appropriate or fair way to dispense medical resources?

When a new medical discovery or advance is made that may help to relieve the symptoms of aging or extend life, the news media often overpromise what that particular advance can do. Science advances incrementally, but media professionals untrained in science can overstate the significance of particular findings. Scientists themselves, untrained in communicating with lay audiences, can mislead the public. All of this is unintentional, but it has an effect. And sometimes attempts to mislead are deliberate. Marcon, Murdoch, and Caulfield (2017) examined 214 articles from 2015 to 2016 using the keywords "stem cell(s)" on websites known to contain junk science. The claims they found were often misleading or exaggerated and seemed to promote mistrust of science and medicine.

Many such claims promise that the user of stem-cell-related treatments will "achieve a higher life quality" or "change your life in 30 minutes." But consumers have little information to go on with these products. Unfortunately, they probably also do not know the status of treatments involving stem cells. Stem cells do indeed promise a revolution in biology and medicine. Some aspects of that promise are being realized, especially in research. The modeling of disease processes and the testing of potential drugs have used stem cells and induced pluripotent stem (iPS) cells to great advantage. However, clinical tests of actual therapies have only just begun.

Lau et al. (2008) examined a number of these treatments that were offered directly to consumers via the internet. Treatments included those for multiple sclerosis, stroke, Parkinson's disease, spinal cord injury, Alzheimer's disease, ischemic heart disease, cerebral palsy, autism, and Duchenne muscular dystrophy. The types of stem cells in the treatments included adult autologous stem cells, fetal stem cells, cord blood stem cells, and embryonic stem cells. Cells were infused at multiple sites. Essentially none of these treatments had any peer-reviewed scientific support, and all overpromised effectiveness. In the decade since their study, the problem has continued to grow.

Research into Aging: General Arguments

Progress in biomedical research in the last decades has been astonishing. Advances that would have seemed like science fiction only a few years ago are commonplace now. Skin cells have been transformed into stem cells. The immune system has been coaxed to attack cancer cells. Amazing new imaging techniques have been developed. The result of those and many other advancements has been the continuing extension of human life span. At first glance, extending the life span of humans sounds like a great idea. Who would not be happy to get extra years of life? The choice would be even easier if those years were healthy ones, free of dementia, frailty, and decline in mental facilities. However, extending human life span is also likely to have significant and even unforeseen consequences for humankind, and so it is the subject of considerable ethical debate.

As a result, a number of scientists, ethicists, and others have concerns about that research. Some caution that a go-slow approach is desirable, and others actually oppose the research. The concerns about the research tend to fall into one of two sets: the research in general and specific aspects of the research. Of course, these positions are not mutually exclusive, but they have led to serious discussion about the value and appropriateness of the research. In some cases, objections to scientific research are based on political reasons or religious reasons (e.g., research involving material obtained from aborted fetuses or stem cells).

The quote from Chekhov that begins this chapter sums up the feelings of those who generally disapprove of research to extend the human life span. Many of them see that research as interfering with the natural order. These arguments have a significant religious component. Those who make this argument tend to be conservative and to oppose abortion and stem cell research. Advocates for the research point out that humans have been interfering with the natural order of things for thousands of years. Easy examples are crossbreeding to develop better crops and animals and research to improve health care. The advocates ask how these improvements are different from seeking to extend life span.

People living longer lives might greatly affect many social systems, including employment, health systems, resource allocations, retirement, and more. Opponents imagine a world of much older people with nothing to do except either block younger workers from employment openings or wander aimlessly. On these and other grounds, the President's Council on Bioethics (2003) argues for caution in research to extend human life span. Its passionate and elegant defense of the natural order of life is well worth reading:

> Without some connection between change and permanence, time and the eternal, it is at best an open question whether life could be anything but a process without purpose, a circumscribed project of purely private significance. Our natural desires, focused on ourselves, would lead us either to attempt to extend time as far as technologically possible or to dissolve it in the involution of a ceaseless series of self-indulgent distractions.

However, although the language is beautiful and the essay makes interesting points, overall this passage seems to argue against scientific progress. The main counterargument might be that humans have already extended life expectancy considerably. Furthermore, most biomedical research will continue, despite some scientists urging caution. The current unauthorized experiments with stem cells in humans are a great example. It's hard to put the cat back into the bag.

Some have labeled these "go slow" or "no go" arguments as the precautionary principle (Harris and Holm 2002). The suggestion is that science should tread carefully for fear of unforeseen consequences in attempting to extend life span. The same argument has previously been applied to environmental issues, GMOs, and public health. It says that those who propose an activity to fulfill some perceived need should have the burden of showing that their suggested activity will not cause harm. Harris and Holm go on to suggest that "when an activity raises threats of serious or irreversible harm to human health or the environment, precautionary measures which effectively prevent the possibility of harm (e.g., moratorium, prohibition, etc.) shall be taken even if the causal link between the activity and the possible harm has not been proven or the causal link is weak and the harm is unlikely to occur."

Of course, science has examples that support such a position. Thalidomide was a popular drug for morning sickness in the late 1950s, before it was shown to cause serious birth defects. Egas Moniz received the Nobel Prize for developing the lobotomy procedure, before it was used inappropriately as a means of controlling behavior among certain groups of vulnerable people. Rabbits were introduced into Australia for sport, but the extraordinary reproductive capacity of rabbits soon made them a destructive pest. Kudzu was introduced into the southeastern United States as a decorative plant, but it now has taken over large swaths of land, killing trees and other native plants.

Although these and other examples can be cited as failures, counterarguments are possible. Science has been fairly cautious, especially in recent years. Thalidomide was never approved by the FDA for use in the United States, and so the birth defects did not occur in the United States. In the 1970s, scientists voluntarily limited recombinant DNA experiments until they could be proved to be safe. More recently, scientists have agreed to avoid some types of experiments, such as those making changes to the germline. And, of course, one must weigh the effects. Is the harm serious? Does it outweigh the good that might come from the activity?

The precautionary principle is also invoked in discussions of genetic manipulation of the human genome or of human reproduction. UNESCO's International Bioethics Committee insisted that "the human genome must be preserved as common heritage of humanity" (UNESCO 1997). The idea is to leave well enough alone, or, to put it another way, if it ain't broke, don't fix it. But we do not know if a fix now will cause a harm in the future.

Other critics bring up the consequences of people living longer lives. There would likely be social disruptions due to the need for food, jobs, housing, transportation, and medical care. Even now, Social Security and Medicare periodically need adjustments in order to deal with the larger number of older people today. If people began to live longer than 100 years, such programs would be stretched even more. More extreme scenarios might feature overpopulation, draining of resources, and civil strife. Advocates counter that we have made adjustments as life span extended over the last 100 years, and we can do more. It is true that society has survived the rough doubling of life span over the last 100–200 years and managed to adapt.

Others protest that a longer life with a poor quality of life is not worth the investment. Two relatively recent examples argue against that position (Lucke et al. 2009). First, oral contraception caused relatively sudden and dramatic changes in mores as women could control their reproduction independently. It also allowed women to participate more fully in the workforce. Second, hormone replacement therapy improved the lives of countless postmenopausal women. It extended the health span of many women and quickly became the most prescribed medication in the United States. These two pills influenced life in the United States and elsewhere, but the world was able to adjust to them. This suggests that a treatment that extends life would change lots of things but could be absorbed by society.

Quality of life is another concern. Longer lives would probably be desirable if the extra years allowed for mental and physical health, but few would want an extra twenty years if those years are spent in dementia or bedridden. But what about if there was no significant extension of life, but only a compressed senescence at the end of life? Would that be a desirable outcome (Lucke and Hall 2005)?

Some authors mention the "longevity dividend," which would allow people to work longer, accumulate more wealth, and achieve more of their personal goals. Extending human life may offer economic benefit too. Murphy and Topel (2003) estimated that the benefit of longer life span has amounted to about $2.6 trillion per year since 1970. Averaged across all people, the amounts are huge: an additional 10 percent reduction in deaths from heart disease would yield $4 trillion in gains, and a 1 percent reduction in deaths from cancer would yield about $400 billion. They suggest that these gains make investment in biomedical research imperative.

Bioethical Controversies in Specific Research Areas

Some people (e.g., Pijnenburg and Leget 2007) say extending human life span is undesirable and morally unacceptable. Others see ethical issues not with the attempt to extend life span but with the methods used in aging research, such as stem cells, genomics and big data, and genome editing.

Stem Cells

Although stem cells come in many types, the main objection to their use focuses on embryonic stem cells, which are derived from aborted fetuses (Hyun 2010). Those who oppose abortion also tend to oppose the use of the aborted embryos. To them, the five-day-old preimplantation embryos in a culture dish or a woman's body are ethically equivalent to living people. However, other types of stem cells do not come from embryos, and the discovery of iPS cells by Shinya Yamanaka in 2006 provided a readily available source of pluripotent cells donated by patients.

iPS cells provide scientists with a reliable source of stem cells without the possible moral ambiguity of embryonic stem cells. However, as is often the case, solving one set of problems reveals another set. For example, informed consent for the use of donated cells is far more complicated than first thought (Aalto-Setälä, Conklin, and Lo 2009). Standard consent forms work well for many procedures, but the nature of the work involving cells donated for iPS cells introduces new areas of concern. How will the cells ultimately be used? Who will get them? The authors listed five areas that stretch the notion of informed consent: genetic modification of the cells, injection into nonhuman animals, genome sequencing, sharing lines with others, and patent and royalty issues. Of course, many of these areas are exactly the ones that help to advance research. For example, the sharing of cell lines is critical for different laboratories to repeat experiments and compare results. The number of experimental variables is greatly reduced if two groups can work with the same cell lines. The early versions of consent forms lacked full information about the possible uses of those donated cells. The use of stem cells to generate synthetic embryo-like structures and organoids and their commercialization raise additional ethical issues (Boers, van Delden, and Bredenoord 2019).

Advances in stem cell biology have resulted in laboratory models that are close to indistinguishable from embryos (Rivron et al. 2018). Stem cells in culture spontaneously organize themselves into three-dimensional organoids. In fact, some also form the extra-embryonic tissues that support development. Those models can be implanted where they begin to develop. Clearly, they are outstanding models for studying embryogenesis. However, they also involve serious ethical questions about the status of the developing tissues.

In a fascinating paper, Monika Piotrowska (2019) notes how the advances in stem cell biology have complicated ethical considerations. The transfer of a somatic cell nucleus into a denucleated egg can develop into a complete organism. Providing the correct transcription factors to somatic cells can reprogram them to become induced pluripotent stem cells. More recent experiments have shown that stem cells possess the information needed to form simple structures if given only a confined space. Is there a point at which these developing stem cell organoids should be considered "human"? Most times the answer to that question has been no, because the organoid lacks the potential to become a human. However, Piotrowska argues that as science advances, that might not be the case forever. Where do we draw the line if every egg or somatic cell has the potential to become human? Clearly, the ethical waters are becoming extremely murky.

Big Data and Genomics

Completion of the Human Genome Project and the development of high-speed DNA sequencing techniques have brought genomics to research into aging and even close to clinical application. The combination of genomic information and Big Data is valuable for patient care and for developing new treatments. However, the use of that information brings additional responsibilities for its stewards (Joly et al. 2016). These include intellectual property rights, recognition of work, proper curation of data, and privacy issues over genetic information that is inevitably identifiable. In addition, researchers must obtain consent from the patient and maintain the confidentiality of the information while sharing the data with others (Francis 2014). It's a difficult balance. The Office of Human Research Protection at the National Institutes of Health revised the regulations to protect human subjects (Fisher and Layman 2018). The new regulations offer choices about consent for storage and future use of identifiable data and how consent is obtained for collection, storage, and use of data by investigators other than those associated with the original study. These regulations provide considerably more protection to research participants.

Editing Genomes

The ability to edit genetic sequences in vivo received a great boost with the discovery of CRISPR (Jinek et al. 2012). Using CRISPR and its many variations, gene sequences can be easily and precisely modified, and this ability opens up worlds of possibilities for treating genetic diseases. Critics point to the potential hazards of modifying genes, especially in the germ cells. However, scientists are very sensitive to these issues, and a 2017 paper (Mulvihill et al. 2017) urged caution, especially in human experiments.

CRISPR will affect many areas of research. It could make crops more nutritious, allowing them to feed more people. It may allow new tools to cure diseases. Genes that we know are associated with Huntington's disease, cystic fibrosis, and breast cancer might be modified to prevent those diseases. It might help in the production of new antibiotics and antivirals. Using a tool called gene drive, scientists can change the likelihood that a particular gene will be inherited from 50 percent to almost 100 percent; in that way, they might alter an entire species. For example, ticks might be changed so that they no longer carry Lyme disease.

Critics point out that these types of interventions can have completely unpredictable consequences. While these experiments might sound great in theory, we lack a sufficient understanding of the complexities of nature to be fiddling about with the gene pool in this way. Already the genetic modification of food crops has raised significant concerns, even though no problems have been yet found.

When it comes to gene editing, three particular areas of concern have been noted (Brokowski and Adli 2018). First, there are technical limitations; these include errors due to limited on-target editing efficiency, incomplete editing, and inaccurate on- or off-target editing (Zhang et al. 2015). These limitations may be overcome by improvements to the technology. Second, will the modified organisms be affected indefinitely and will the edited genes will be transferred to future generations? Can accurate risk-benefit analyses be done given how much we do not yet understand? Third, even if the editing is successful, do we really understand how genetics and phenotypes interact, so that we can make accurate predictions about the ultimate outcome? These concerns argue strongly for careful consideration of appropriate guidelines and regulations. In fact, the National Academies (2017) published guidelines for the ethical uses of CRISPR. Scientific organizations in other countries have also.

Aging Research as a Moral Imperative

Aging research has two purposes (Gems 2011). The first is to better understand the mechanisms of aging. The second is to cure diseases of the aged. The arguments used by those who oppose research into life span extension have been used many times in the past to argue for slowing down or avoiding some areas. However, while some might have concerns about research to extend life span, it is very difficult to argue for slowing down research into specific disease areas.

For example, until the 1950s, polio was an annual scourge. Children in braces and iron lungs were a common sight. Similarly, the amazing effort to eradicate smallpox saved millions of lives. Who could argue with those achievements?

Today there are similar groups of people suffering from various diseases—Alzheimer's disease, Parkinson's disease, terminal cancer. Strategies that cure or prevent those diseases would be major accomplishments. Many deaths would be prevented, and lives would be improved. Research into these diseases might be thought of as a moral imperative, even a moral duty (Harris 2005).

15

Future of Death

The job of theorists, especially in biology, is to suggest new experiments. A good theory makes not only predictions, but surprising predictions that then turn out to be true.

— Francis Crick, *What Mad Pursuit: A Personal View of Scientific Discovery*

Scientists make predictions all the time. As Crick notes above, prediction is an essential part of science, and the results can be surprising and valuable. But can we make predictions about the future of death? As we eliminate present-day causes of death, will others appear or increase to replace those that were eliminated? Will changes in our environment put us at new risk? These are all questions without answers. But we can speculate and even, as Crick indicates, make a few predictions.

Our predictions might be informed by extrapolations from human experience over the last few hundred years. Humans have taken some control of disease and death. Over that time, both the average life span and the quality of life have improved significantly, at least in the developed world. Most of the improvements early on came from limiting infant mortality. Interestingly, it was not always some great breakthrough in medical science that resulted in longer and better lives. Many improvements occurred without any real understanding of the connections between the changes and better health. Water supplies and sewage systems were established in the nineteenth century, before the relationship between poor sanitation and infectious disease was known. John Cairns pointed out this fact in his excellent book *Cancer, Science, and Society*: "It is significant, however, that what is thought of as one of the accomplishments of sophisticated medical science was, in large part, the product of some fairly simple improvements in public health" (Cairns 1978, 7–8). He also noted that we do not always have to understand the mechanisms of diseases to make changes that will significantly reduce their morbidity and mortality. In the book, Cairns made the case for ending the environmental causes of cancer (e.g., smoking, air pollution). He asserted that we did not need to know the underlying mechanisms that produced cancer; the strong correlation between cigarette smoke and lung cancer

The Biology of Death. Gary C. Howard, Oxford University Press. © Oxford University Press 2021.
DOI: 10.1093/oso/9780190687724.003.0015

was obvious, and we could take measures to protect people and figure out the exact mechanism later.

During the past two centuries, the ability to prevent and treat infectious diseases has improved dramatically. Infections had been a major cause of death ever since humans evolved. In 1846, Hungarian physician-scientist Ignaz Semmelweis noticed a difference in mortality in two maternity wards in his hospital. The ward overseen by male physicians and medical students had five times more deaths than the one supervised by female midwives. He also noticed that a pathologist friend died of a similar malady. In fact, it was common for pathologists to die that way. Semmelweis concluded that the doctors were picking up something from the cadavers that they worked on. He ordered them all to wash their hands with a chlorine solution before treating patients. The rate of childbed fever dropped dramatically. Unfortunately, his findings were not translated into general practice for many more years. In 1928, Alexander Fleming noticed some peculiar growth patterns in some petri dishes in which he had been growing bacteria. One plate had been contaminated by a fungus, and the bacteria were not growing in the area around the fungus. That fungus was secreting what we now know to be penicillin, and Fleming's discovery revolutionized the treatment of infectious disease with antibiotics. Since then, vaccines have helped to eliminate many viral diseases. Measles, mumps, diphtheria, pertussis, smallpox, and polio have been eliminated or greatly reduced.

Advances in many other areas have also helped. Medical breakthroughs have been accomplished in imaging, chemotherapies, information technology, minimally invasive surgical techniques, and combination therapies for HIV. Legislation has resulted in much cleaner air and water, safer and better food, more effective drugs, less smoking, better public education, and many other improvements in the environment that have translated into longer life spans, as Cairns suggested.

These examples might give us hope that humans are likely to be able to deal with major challenges to life and health in the future. In fact, the rate of mortality in the United States has decreased over the past 100 years and is expected to continue to drop. Average life expectancy of Americans will reach the mideighties by 2070 and creep up to the late eighties by 2090 (Penn Wharton 2016). These are averages over all groups. Not all groups will fare so well. For example, African Americans and Hispanics have always had lower life spans than white Americans, and in recent years, lower-class whites with less than a college education have suffered a reduction in life span.

Not everyone agrees about the possibility of extending human life span. Barbi et al. (2018) examined the cases of 3,836 Italians aged 105 and older and discovered that the risk of death after age 105 was essentially constant. This is an unusual result, because in general, as we age, the risk of death increases each year. The researchers referred to this result as the "plateau of mortality," and it suggests that there is no limit to human life span. However, their findings are controversial. Others suggest that life span is limited to about what it is now (Beltrán-Sánchez, Austad, and Finch 2018), and still others (Newman 2018) blamed the plateau on statistical errors.

Pijnenburg and Lege (2007) argue against extending life on three grounds. First, they point to the moral problem of unequal death. How could the ability to extend life be fairly offered to both the rich and poor? If it cannot, it should not be offered to anyone. Second, they note the importance of community to a full human life. We live not as individuals but within a community of our fellow humans. If we cannot do that with extended life, then such a life would not be worth living. They think we should be focusing our attention on improving the lives of all rather than trying to extend life for no good reason. Finally, extending life would violate many religious and other spiritual traditions. Most of those traditions teach us not to dwell on ourselves but to make life better for others. Expending life is not consistent with these teachings.

New Technologies

Public health and medicine have made great strides to extend life span and health span. There is no reason to believe that this progress will be less successful in the future. Several areas of research have been remarkably productive in recent years and promise to extend human life. One wonders what other breakthroughs will occur in the next 100 years.

Stem Cell Biology and Regenerative Medicine

"Imagine with me a world where for so many human conditions, spinal cord injuries, Parkinson's, diabetes, and heart disease, that we are somehow able to create cells in your organs to get to the root of the problem and actually restore that organ's function. That is the world we are heading into" (Srivastava 2015).

In the Middle Ages, alchemists sought the philosopher's stone, which they believed could turn base metals such as lead into gold. It could also cure disease and extend life if dissolved in solution, creating the elixir of life. The alchemists searched for it in vain, but modern cell biologists may have found it in stem cells (Triffitt 2002). Since they were discovered in the 1970s, stem cells have tantalized medical scientists with the possibility of repairing damaged or diseased organs. Stem cells are defined by two properties: they can divide indefinitely to create more stem cells, and they can differentiate into any cell type in the body.

These two properties make them enormously interesting as a potential solution for many diseases involving damaged or dead cells and tissues. Since they can develop into any adult cell type, they might be used to repair hearts after heart attacks, brains after strokes, spinal cords after injury, or a pancreas that no longer secretes insulin. Already experiments are under way to use stem cells to treat macular degeneration and other diseases.

However, before the promise of regenerative medicine can be realized, several fundamental problems must be solved. Can a sufficient number of cells be generated? Can the cells be prepared in a way that avoids creating new problems? Can they be transported to the correct location in the body? Will they incorporate in a way that allows them to function appropriately?

The field gained a huge advance in 2006, when Takahashi and Yamanaka reported that by using just four proteins—transcription factors that turn genes on or off—they had "reprogrammed" mouse skin cells into a type of stem cell called induced pluripotent stem (iPS) cells (Takahashi and Yamanaka 2006). A year later, Yamanaka reported that they had also succeeded with human skin cells (Takahashi et al. 2007). Their work was quickly replicated in many laboratories around the world. With iPS cell technology, scientists suddenly had a ready supply of stem cells that could be directed to develop into any desired cell type. More importantly, Yamanaka's work also showed that the developmental state of cells was not fixed. For some time, it had been assumed that embryonic stem cells divided and then took on a specific developmental path to become brain, heart, liver, skin, or other adult cell types. The path was assumed to be one way, embryo to adult. Yamanaka showed that it is a two-way path.

Other scientists took Yamanaka's work to the next step and showed that any adult cell can be reprogrammed to another adult cell type without going through the intermediate stage of becoming a stem cell if the correct set of transcription factors is used. Thus, the state of a cell seems to be completely fluid. For example, Deepak Srivastava (Qian et al. 2012) used a different set of transcription factors to reprogram fibroblasts in a mouse heart directly into cardiomyocytes, the beating muscle cells of the heart.

Yamanaka had found the philosopher's stone for stem cells and ushered in a new era in biology. iPS cells have been used to study cell development, as a model for various diseases, and to test new drug therapies. More recently, researchers have begun to explore their use in regenerative medicine, and it is likely that, in the future, this work will help physicians to repair or replace damaged organs and extend human life.

While most experiments with reprogramming have concerned the repair of damaged tissues and organs, other studies have looked at using the same reprogramming factors to rejuvenate cells (reviewed in Goya et al. 2018). For example, one group found that a six-factor cocktail resulted in a higher doubling potential, longer telomeres, and more effective mitochondria—in essence, resetting the aging clock. And some studies have looked at whether reprogramming cells might show promise in treating macular degeneration, diabetes, and Parkinson's disease.

The Human Genome and Genome Editing

At the turn of the century, Francis Collins and Victor McKusick predicted that molecular genetics would become a routine part of clinical practice: "The power of the molecular genetics approach for answering questions in the research laboratory will catalyze a similar transformation of clinical medicine, although this will come gradually over the course of the next 25 years" (Collins and McKusick 2001).

In recent years, several techniques for gene manipulation have been developed, including zinc finger proteins, TALENS, and meganucleases. Those worked, but the real advance has come with the addition of CRISPR editing to the toolbox. The CRISPR-Cas system uses a guide RNA to specifically target the nuclease to the exact place in the sequence to make a double-stranded cut. Then the host cell's own repair mechanism takes over to fix the breaks. The system is based on one used by microbes. The microbes capture a small sequence of a pathogen's DNA and use that sequence to look for other copies of the pathogen DNA. When they find them, they cut up the invading DNA to stop the infection. That sequence is then used to target the nuclease component of CRISPR to destroy the invader. Jennifer Doudna and Emmanuelle Charpentier won the 2020 Nobel Prize in chemistry for the discovery of CRISPR.

Now scientists are using this system and numerous variations on it to edit DNA genomes with great precision (Knott and Doudna 2018), adding or delete small sequences. This ability promises to allow physicians to treat genetic diseases by "genome surgery," a term originated by Bruce R. Conklin. As one example of the power of this technique, Conklin and Doudna plan to use genome editing to treat Best disease, a disease that involves a dominant mutation that causes blindness. With the mutant gene edited out, the recessive (normal) allele will be expressed and sight may be restored. They selected this disease because the eye is accessible, a relatively small number of cells can make a significant difference, and the dominant gene is vulnerable to a correction by CRISPR. These experiments will provide an excellent demonstration of the power of this technique, but it may take more time to find treatments for other genetic diseases.

In terms of basic research, CRISPR has been used to insert fluorescent markers into DNA to examine the localization and expression of individual genes in single cells. This is critical because subtle changes in gene expression have major effects on regulation. The ability to monitor gene expression at this level is revolutionary.

Efforts are now focused on multiplexing the actions of CRISPR (Tachibana 2019)—that is, editing more than one gene at a time. This would greatly accelerate developing T-cell-based immunotherapies that require changes in multiple genes. CRISPR might also be used to create "stacks" of genes that are inherited together. This would be beneficial to plant scientists so that they could introduce genes with resistance against multiple pests at the same time. One goal of the research is to eliminate the necessity for any nucleic acid. The ability to introduce only proteins into cells would eliminate the possibility of incorporation of genetic material into the genome.

Immunotherapy and Other Cancer Therapies

Even after decades of research, cancer remains one of the major causes of death (see Chapter 3). Recent advances in understanding our immune system have resulted in exciting new treatments. Our immune system protects us against a myriad of invaders, such as viruses and bacteria, and at the same time it can recognize our own healthy tissues and does not attack them. For years, scientists wanted to know how to unleash this powerful defensive system against cancers. Cancers are our own cells that have lost their ability to control their growth. If the immune system could detect them as an enemy, it might be able to destroy the tumor.

To do its job properly, the immune system must be highly regulated. It has many opposing systems that turn it on or turn it off, and that balance must be carefully maintained (Pardoll 2012). Among those systems are what are called checkpoints or inhibitory pathways that help to modulate the immune response.

James P. Allison and Tasuku Honjo received the 2018 Nobel Prize in medicine for immunology-related discoveries. Allison discovered a protein called CTLA-4 that inhibited the immune system at an immune checkpoint (Leach, Krummel, and Allison 1996). He made an antibody against the protein and found that it prevented cancers in mice. Honjo discovered another protein, called PD-1, that does something similar, albeit by a different mechanism (Nishimura et al. 1999). Their discoveries have added another powerful weapon against certain cancers. For example, former President Jimmy Carter was successfully treated for advanced melanoma with an antibody drug against PD-1.

Still more technologies are attacking cancers. Chemotherapies may kill 95 percent of cancer cells, but the remaining 5 percent survive to cause a recurrence of cancer or metastasis. The ability to look at very small samples will allow scientists to determine what is different about that last 5 percent and to learn how to kill those cells too. Other scientists at Georgia Tech have developed magnetic nanomachines about the size of ribosomes (small (20–30 nm) organelles of RNA and proteins that make proteins) that can degrade tumor cells. They also may be used to open clogged arteries in the heart.

Personalized Medicine

All of these technologies and others will allow physicians to diagnose and treat patients much more as individuals (Harvey et al. 2012). Patients will be examined at the molecular level to better understand their exact genetics and predispositions for specific diseases. Furthermore, therapies will be tailored to meet their exact needs. Disease progression will be followed, and the effectiveness of specific drugs will be monitored with much greater precision. Improved data collection and analysis will make the latest therapies more available to patients who need them.

New Threats

Based on the historical record and these and other medical breakthroughs, one might feel fairly confident in predicting that human life span would continue to increase and that the quality of life of older people would likewise be improved. Yet even with all of this progress, there is reason for concern. In part, this is because the most serious threats facing us are not necessarily connected to disease. For example, weapons available now are far more powerful than the ones used in previous wars, and more nations and non-nation groups have or are trying to get them. Beyond that, global warming has the potential to disrupt civilization on a massive scale. It is by far the greatest threat to humans.

Influenza Redux

The influenza pandemic of 1918 killed more than 50 million people. The particular combination of mutations that made it so lethal has not occurred since, but it (or something similar) could. Influenza A mutates rapidly. These viruses contain two proteins that determine their infectivity and virulence. There are 18 types of hemagglutinin (H) and 11 types of neuraminidase (N). The different combinations of these proteins are why we need a new flu vaccination every year. The concern is that the 1918 strain or something equally infectious could emerge. Three flu pandemics have occurred since 1918: in 1957 (H2N2), 1968 (H3N2),

230 THE BIOLOGY OF DEATH

and 2009 (H1N1). These pandemics happen when the normal circulating strain of influenza A is replaced by a new, very different strain for which humans have little immunity. Avian influenza A strains mostly infect birds, but some are of special concern because of their exceptional potential to infect humans and to spread rapidly. Some are much more virulent than the "typical" influenza strains. For example, the H5 viruses cause 50 percent mortality, and the H7N9 strain caused 40 percent mortality in humans.

Emerging Diseases

Emerging and reemerging viruses are an ever-present problem (Gao 2018). Outbreaks of Ebola (a filovirus) have occurred in Africa in the last several years. This often-fatal disease results in bleeding from many parts of the body, and it has struck terror in those areas. Zika virus is not often fatal, but causes deformities in newborns. Other emerging viruses include other filoviruses (Marburg), henipaviruses (Nipah, Hendra), Lassa virus, Lujo virus, South American hemorrhagic fever viruses (Junin, Machupo, Guanarito, Chapare, Sabia), Crimean-Congo hemorrhagic fever virus, Rift Valley fever virus, hantaviruses, SARS coronavirus, MERS coronavirus, and tick-borne encephalitis viruses.

In early 2020, a novel coronavirus, SARS-CoV-2, swept around the world. While many of those infected were asymptomatic, others and particularly those with preexisting conditions (e.g., lung disease, heart disease, diabetes, immunosuppression) had a much worse course. Large numbers of people died, and there was significant economic disruption. The source of the virus is not certain, but it is believed that it crossed from bats into humans in China (Andersen et al. 2020). While it has tragically claimed many lives, it could have been much worse if the virus had been as virulent as, for example, the 1918 influenza. Nevertheless, more than 400,000 Americans died of Covid-19 in less than a year.

The incidence of vector-borne diseases has increased in recent decades, both from endemic diseases and from vectors that are enlarging their area (Kilpatrick and Randolph 2012). Global warming presents another factor to consider in emerging diseases as vectors move into new areas. For example, climate change has allowed a significant expansion of the range of the mosquito *Aedes albopictus*, which bites during the day. This mosquito is a public health concern because it is a vector for many arboviruses, including the virus that causes dengue fever. *A. albopictus* and *A. aegypti* are also associated with epidemics of chikungunya, another viral pathogen (Lambrechts, Scott, and Gubler 2010).

The increasing urbanization of humans is another key factor (Alirol et al. 2010). One hundred years ago, only 20 percent of the world's population lived in cities. Now half does. In most developed nations, this migration has resulted in improved education, healthcare, and quality of life. That is not the case in many underdeveloped countries. There many of the residents live in shantytowns with little access to clean water, sanitation, and medical care. For example, in 2009, in the Central African Republic overall, 96 percent of the people living in urban areas lived in slums; in Nairobi, Kenya, the number was 60 percent (Neiderud 2015). Slum residents are at increased risk of infectious diseases. For example, poor sanitation means that human waste builds up in those shantytowns, and this increases the risk of diarrheal diseases. In addition, vectors accumulate in those areas. Rat populations explode. Rats carry many pathogens, including *Yersinia pestis* (plague), *Leptospira*, *Rickettsia typhi* (typhus), *Streptobacillus moniliformis* (rat bite fever), *Bartonella*, Seoul hantavirus, and *Angiostrogylus cantonensis* (rat lungworm). Mosquitoes that carry dengue, yellow fever, and chikungunya breed easily in water that accumulates in discarded containers or tires. Diseases that were formerly exclusively seen in rural areas, such as Chagas disease and leishmaniasis, have migrated to urban areas as well. And increased movement of people around the world means that those diseases can spread more rapidly and appear in places where they have not typically been seen.

However, not all of the increased risk of disease can be laid on vectors. The movement of people to cities also resulted in changes in cultural norms and behaviors. One result has been an increase in the incidence of sexually transmitted diseases, and one of those diseases, HIV/AIDS, is one of the most lethal pandemics in history. We lack both immunity and effective drugs against novel viruses, and available antibiotics are becoming less effective as bacteria develop resistance to them. A whole new set of environmental challenges are on the horizon. New social and environmental forces, including economic downturns, climate change, and human migrations from the Middle East and North Africa, are merging into a "perfect storm" to promote the widespread emergence of neglected tropical diseases (NTDs) in Southern Europe (Hotez 2016). Many more zoonotic viruses (viruses that jump from animals to humans) are waiting in the wings for their chance to cause outbreaks, epidemics, and pandemics: Ebola, Marburg, Sin Nombre, simian foamy virus, monkeypox, Junin virus, Vachupo virus, Lassa virus, Hendra virus, and many more as yet unnamed.

Antibiotic Resistance and Bacterial Infections

For the last several decades, antibiotics have been miracle drugs, helping to overcome the threat of bacterial infections. Unfortunately, we have not always used antibiotics wisely. Overuse and misuse of these drugs have allowed bacteria to develop resistance to many of them, and some show resistance to multiple antibiotics (Levy and Marshall 2004). Even worse, for years, the research and development pipeline for new antibiotics had slowed to a trickle. Thus, in the future, we may have fewer options for treating bacterial infections. The situation is serious. In fact, the World Economic Forum recently reported that "arguably the greatest risk . . . to human health comes in the form of antibiotic-resistant bacteria. We live in a bacterial world where we will never be able to stay ahead of the mutation curve. A test of our resilience is how far behind the curve we allow ourselves to fall" (Howell 2013).

Antibiotic resistance is not a new phenomenon. Bacteria and yeast have been making antibiotic molecules throughout evolutionary history. We only joined this "arms race" recently. Antibiotic resistance has been found among bacteria deep underground that have been isolated from the surface for millions of years. Spellberg, Bartlett, and Gilbert (2013) suggest a number of strategies to overcome this problem: controlling new infections, encouraging the development of new antibiotics, preserving current antibiotics by using them optimally, developing new types of treatments for bacterial infections, and developing treatments that target the host rather than the pathogen. One measure that would certainly help would be to stop the use of antibiotics in animal feed.

We have taken antibiotics for granted for decades. We always assumed that there would invariably be one more that physicians could reach for as resistance developed in various strains. That may no longer be the case. Without a change in our current practices, we face a future in which bacterial and other infections might again increase to become a threat and a significant cause of death for humans. A large-scale international effort is needed to take on the multiple facets of this problem. (Laxminarayan et al. 2013).

There is some good news. After years of little progress, now new antibiotics are in development. Some are conceptually new, and others are variations on an old theme. Interestingly, some drugs are being developed to enable older antibiotics to work even after many bacteria are resistant to them. For example, bacteria have developed resistance to cefpodoxime, a broad-spectrum antibiotic, by opening its beta-lactam ring. However, various compounds have been developed that inhibit the beta-lactamases and render cefpodoxime active again (Bhattacharjee et al. 2008).

Neurodegenerative Diseases

Neurodegenerative diseases are a major public health crisis in the United States and the world, and the crisis is increasing in scope. While deaths from heart disease and several types of cancers have been decreasing, neurodegenerative disease is becoming more prevalent. Today, more than 5 million Americans are suffering from Alzheimer's disease (Alzheimer's Association 2016). It's the sixth-leading cause of death. By 2050, that number is estimated to increase to nearly 14 million as the baby boomer generation ages. The other major neuro-degenerative diseases, such as Parkinson's disease, amyotrophic lateral sclerosis, and frontotemporal dementia, are also increasing.

The outlook is bleak, as no disease-modifying drugs are available for any of the major neurodegenerative diseases. The few drugs out there only work partially for some patients for a while. Worse still, the search for drugs to treat these diseases has slowed considerably, as major drug companies are reluctant to undertake the expensive testing. Few reliable models of the disease exist. Theories are hard to test because symptoms of the disease take so long to emerge. And clinical tests based on animal studies have yielded poor results.

Diseases of Plenty

The major advances in public health over the last centuries and medical advances (e.g., vaccines, antibiotics) have pushed infectious diseases out of the top spot for causes of death. Chronic diseases are now the major public health threat in essentially all countries (Choi et al. 2005; Choi et al. 2008). These are not communicable in the traditional sense: no pathogen passes from one person to another. However, they are highly "communicable" in a new sense: these are diseases of lifestyle, and lifestyle can be passed along from one generation to the next.

For millennia, humans struggled to find enough food. Our hunter-gatherer ancestors spent most of their time searching for food. Once humans settled into villages and began agriculture, they still depended on enough rain and good weather to ensure a successful crop. Drought, floods, and famine were omni-present threats.

Today we have controlled most of those issues. We have the opposite problem: a world awash in calories. For most of us in the developed world, food is plentiful, even too plentiful. Easy access to prepared food in supermarkets, fast food, and oversized portions in restaurants has changed our eating habits. Everything comes with cheese. Now we can easily take in too many calories in a day. In fact, the struggle is to control our intake of calories. Worse still, our system is still evolutionarily programmed to make the most of every calorie. That system was critical for millennia. Today it works against us.

In the twenty-second century, diseases of affluence will be the major cause of death (Choi et al. 2005). The causes include urbanization, processed food, obesity, stress, pollution, and a sedentary lifestyle. As a result, obesity and its complications, including diabetes, heart disease, and cancer, are growing problems. And too many people self-medicate with alcohol, tobacco, drugs, and food.

In recent years, public health officials noted a striking rise in the number of deaths among primarily white, non-Hispanic men and women with little education (Case and Deaton 2015). This trend is counter to those of all other groups. In those groups, mortality is decreasing. In fact, no other country has reported such a rise. The increase involves death from drugs and alcohol and suicide and occurs in midlife. It seems to accompany depression and desperation from the crumbling economic opportunities and hopes of that group, and it is associated with declining physical health, declining mental health, and a reduced ability to complete the daily activities of normal living. Furthermore, many of that group report chronic pain and an inability to work.

Climate Crisis

The future of death—whether human life span can be extended—is not clear. Our history is encouraging. New advances in medicine and public health are encouraging and even exciting. However, it seems that we take at least one step backward for every two steps forward. New diseases and threats have emerged as traditional ones have been overcome. Now we face the existential threat of global warming. As John Cairns said about cancer in the 1970s, we already know how to prevent it. Today we know how to stop even global warming. The question now, as it was then, is whether we have the will to do what it will take.

Some scientists have suggested that a potential solution to climate change would mimic the actions of volcanoes that inject particles into the atmosphere. There are multiple examples in history of very large eruptions that resulted in a cooler climate for a number of years from the reflection of sunlight off of the particles and back into space. We could do a similar thing by injecting sulfate particles into the atmosphere. The particles do not remain in the air indefinitely, and so the process would have to be repeated regularly. But it might buy some time for us to come up with a more permanent solution.

Nothing is free, however, and this geoengineering could have a downside. First, while it might give us time to find a real solution, we might simply squander that time by not taking the warming of the planet seriously. The particles might also change our weather patterns and cause ecological problems. And while the planet might get cooler, this would not help with the acidification of the oceans that results from increased levels of CO_2 in the atmosphere. Finally, if we did nothing to slow the increase of CO_2 and methane in the atmosphere and then quit injecting the particles, we might experience very rapid temperature increases.

Global warming will bring much higher temperatures, and at some point those increases will be beyond the ability of humans to survive. Mora et al.

(2017) examined the effects of extreme global heat on human populations. They surveyed all reported heat events from 1980 to 2014 that yielded 783 cases of excess death in 164 cities in 36 countries. From these studies, they calculated a threshold of temperature and humidity that is typically fatal to humans. Amazingly, about 30 percent of the world's population experiences twenty days above that threshold already, and that number will increase to 74 percent by 2100 if nothing is done.

The most direct result of climate change is the effect on human mortality as the ambient temperature rises. Hajat et al. (2014) did this for the United Kingdom. They estimated the number of deaths that will occur in the future as the Earth warms to unprecedented levels. They found that heat-related deaths will increase by 257 percent by the 2050s and by 535 percent by the 2080s. The elderly would be among the populations most susceptible to death from the increased heat.

Global warming will bring hazards other than heat. Matthews, Wilby, and Murphy (2019) predict that heat will be combined with many more cyclones or hurricanes. As an example, they use the 1991 cyclone that devastated Bangladesh as a model. As the world warms, extreme weather events will become common. Crop yields will be affected. So will health problems, including mental health. Infectious disease vectors (ticks and malaria mosquitoes) will migrate to higher altitudes. Ironically, some health issues may be improved (McMichael and Lindgren 2011). As winters become milder, the stress on people's systems may be reduced, and the number of heart attacks and strokes may lessen. As some areas warm, mosquitos carrying disease may enter, but other areas that already had a warm climate might become too hot for mosquitos, and the risk of malaria and other diseases in those areas might lessen.

Access to water is already a problem in many parts of the world, but as the Earth warms, water will become scarcer, and it will become a greater and greater source of national and ethnic conflict (Kreamer 2013). Like water, food is also scarce in parts of the world, and climate change will make that situation much worse (Wheeler and von Braun 2013). The need for food and water will cause desperate populations to try to move to areas with those resources, and that will add to competition for those resources, resulting in serious conflicts.

The World Economic Forum Risk Report (2019) noted, "Of all risks, it is in relation to the environment that the world is most clearly sleepwalking into catastrophe." The Intergovernmental Panel on Climate Change (IPCC) bluntly said in October 2018 that we have at most until 2030 to make the drastic and unprecedented changes needed to prevent average global temperatures from rising beyond the Paris Agreement's 1.5°C target. In the United States, the Fourth National Climate Assessment warned in November 2018 that average global temperatures could rise by 5°C by the end of the century if emissions are not reduced significantly. Extreme weather, loss of biodiversity, pollution, and failure to adapt to climate change and the migration that will result are all beginning and will get far worse. Crops, animals, and ultimately people will suffer and die. Tragically, governments have been slow to recognize the threat. Holding the total average warming at 1.5° or 2.0°C will be difficult, and the outlook is bleak.

Have We Reached the Limit of Human Life Span?

Although humans have conquered many diseases with vaccines, drugs, and other treatments, they have yet to stop aging. Some biomedical researchers believe that aging itself is a disease and that its manifestations are cancer, heart disease, and more. The term "adult (geriatric) failure to thrive" has been used to describe this situation (Gavrilov and Garilova 2003).

The twentieth century saw great gains in life span in many countries. In the first half of the century, the gains were mostly due to a reduction in mortality at younger ages. In the second half, the improvements occurred mostly in older people. Vaupel (2003) argued that the rate of increase in life span has been steady for some time and that there is no reason to believe that it will change in the future. In other words, he believes that human life span will increase significantly beyond what it is today. Those over age sixty will soon constitute 20 percent of the world's population (Hansen and Kennedy 2016). However, it remains to be seen if longer life span can be translated into a longer health span, rather than simply extending the time people spend in failing health. As we've noted in earlier chapters, one key observation has been that many of the features of aging underlie the chronic diseases of the elderly. This presents intriguing questions: Is aging itself a disease? Could those diseases be mitigated by attacking the aging process?

The life expectancy of certain groups has risen faster than that of other groups. The populations of a set of vanguard countries (e.g., Finland, Norway, and Sweden) have longer lives on average than the populations of other countries, presumably because their healthcare system allows better access to high-quality care than is the case elsewhere. Even within these vanguard countries, groups differ in their mortality rates. Those with better access to health care live longer than those who do not. Other factors might also be involved, such as nutrition, education, and sanitation. However, groups that previously did not have these advantages catch up in terms of life span once they have access to them. Jasilionis et al. (2012) argue that these examples indicate human life span can continue to be lengthened in other regions of the world.

Oeppen and Vaupel (2002) disagree that human life span has hit the wall. They make three arguments. First, over the years many experts have claimed that human life span is limited, and they have all been proved wrong. Second, what looks like leveling off is an artifact; the average rate of increase in life span has slowed in some countries, but others are catching up. Finally, the rate of increase should be leveling off if a maximum is being approached. It is not. Therefore, they conclude that there is no reason to believe that humans cannot live much longer.

Most of the calculations of human life span rely on a single model. King and Soneji (2011) instead used a Bayesian approach that took into account various factors, including a reduction in cigarette smoking and an increase in obesity. They estimated that life span would increase and that it might increase faster for men than women. They noted that other factors, such as immigration, marriage, and education, might also affect the estimates. Kontis et al. (2017) also used a Bayesian approach to examine life span in thirty-five nations. They found that life span is likely to continue to rise throughout the developed world. For example, the life span for women will probably be over ninety by the year 2030. Some believe human life span will rise to 125 years (de Beer, Bardoutsos, and Janssen 2017).

Not everyone is convinced. Carnes, Olshansky, and Grahn (2003) compared the life spans of mice, dogs, and humans in mathematical models to see if there was a maximum for human life span. They noted some commonalities. First, the physiological decline due to aging begins at about one-third of the total life span of the animal. Second, humans lose 80 percent of their physical abilities by age eighty. Third, the pathologies seen at death can differentiate young and old individuals. By this, they mean that age takes a significant toll on bodies, regardless of species. They conclude that these observations point to a maximum age limit for all living organisms, including humans.

Olshansky et al. (2005) also saw it differently. While they hope they are wrong, they believe the consistent rise in life span over the last hundred years will slow down in the twenty-first century. Obesity, diabetes, and heart disease are on the rise. Diabetes includes its complications of heart disease, stroke, and kidney failure. Obesity will increase the incidence of heart disease, hypertension, and cancer. Our modern lifestyle with its lack of exercise and poor diet will offset any gains from medical advances and slow the increase in human life span.

Vijg and Le Bourg (2017) agree that the end of war and famine in developed nations has lengthened life span, and that advances in modern medicine have contributed. However, they argue that essentially all animals that reproduce sexually have a relatively defined life span. By analyzing global demographic data, Dong, Milholland, and Vijg (2016) found that the factors that enhanced life span declined after age 100. They believe that is why the age of the world's oldest person has not increased for some time. Based on these results, they suggest that there is a maximum age for all humans that will never be surpassed.

Much of the research over the years has looked at diseases individually. However, there are commonalities in the pathologies of many chronic diseases. Aging is the major risk factor for many of these diseases, including neurodegenerative diseases, metabolic syndrome, many cancers, and cardiovascular disease. One process that is getting more attention in recent years as we look for links between aging and disease is inflammation. Inflammation is a critical part of our defense against disease, but it is also a major factor in both aging and several diseases of aging. Pro- and anti-inflammatory responses exist in a careful balance. As an organism ages, misplaced or damaged molecules are dealt with less effectively by the proteasomes and autophagy. Over time, they accumulate and encourage inflammation (Franceschi et al. 2017).

Epigenetics is another possible common point between disease and aging (Pal and Tyler 2016). Epigenetic marks are thought to lock in specific developmental patterns, but reprogramming of cells has shown that these marks can be reversed. Some evidence suggests that dysregulation of epigenetics underlies the process of aging (López-León and Goya 2017). Some of those changes drive aging, while others are simply associated with aging. Further experiments will determine which ones are which. This is an intriguing possibility that deserves further study.

Aging as a Disease

What is a disease is not always clear. Throughout history, diseases have changed, and diseases have been complicated by historical, societal, and religious influences. Homosexuality was considered a disease as late as the mid-1970s. Disease might be defined as an abnormal bodily function or process that is not caused by injury. That definition makes it easy to describe an infection, such as influenza or typhoid, as a disease. Also, by that definition, an organ failure, such as an inability to secrete insulin in diabetes, would also qualify as a disease.

The definition depends on the interpretation of the term "abnormal." Parkinson's disease involves abnormal loss of control of movement. The common cold involves a series of abnormal symptoms, such as a runny nose, sore throat, coughing, and headache. They are abnormal because of the loss of an ability or the appearance of symptoms, and they have a cause that is well known.

As we've seen, age is a significant risk factor for a host of diseases, including heart disease, cancer, and various neurodegenerative diseases (Niccoli and Partridge 2012). Age also brings other conditions, such as muscle wastage (sarcopenia), loss of bone mass (osteoporosis), hypertension, and dementia. All of these are recognized as disease conditions that warrant medical intervention.

In fact, aging is the most important risk factor for some diseases (Kennedy et al. 2014), and therapies that treat many chronic diseases also extend life span. The association of age with these disease conditions has led some researchers to wonder if aging itself might be a disease. Aging involves a decline in various biological processes, but are those losses of function a disease? To some, those losses of ability are a natural process—as people get older, things go wrong. But others see the differentiation of conditions by age as artificial (Bulterijs et al. 2015). High blood pressure in a thirty-year-old man would definitely be considered a disease. Is it less so for a seventy-year-old man? Amyloidosis is common in older people. The buildup of amyloid in tissues and organs can affect their form and function and can lead to life-threatening conditions. Is that a natural process? Bulterijs et al. (2015) reject this notion. They point out that aging nicely fits the definition of a disease. It involves abnormal functions, and each of its manifestations has a specific cause.

Caplan agrees. He notes that the decline in function actually begins in the thirties, not in old age. He summarized the problem (Caplan 2015) in two questions: Does aging have a purpose, and is it different from pathology? As we have seen here, it is difficult to separate aging and pathology. Gems (2015), in an excellent review, summarizes and then dismisses traditional arguments that senescence is a natural and unavoidable process. Zhavoronkov and Bhullar (2015) believe that there are practical advantages to defining aging as a disease, allowing the biomedical community to tackle aging in a more organized manner.

Predictions Unverified

The future of death is very hard to predict. Some aspects are clear. Humans have certainly made great progress in extending life. Many diseases have been conquered. However, death might be thought of as like the Hydra of Greek mythology: whenever one head was cut off, another grew in its place. In like manner, death seems to grow a new head every time one is cut off. Antibiotics, modern sanitation, safer childbirth, reducing the amount of smoking, cleaner air, and more—all represent death's Hydra heads that have been severed. But as life span lengthened, new heads appeared, such as neurodegenerative disease, heart disease, and cancer. Other, even more dangerous ones lurk nearby: climate change, overpopulation, nuclear weapons. As the second of his twelve labors, Hercules had to kill the Hydra. He did this by cutting off its one immortal head in the center. If aging is a disease, is there one immortal head or regulatory system that can be controlled to stop aging? Will humans be able to kill death? That is too difficult to predict right now.

16

Death Is More than Dying

Death is intimately intertwined with life. Death evolved with life some 3.5 billion years ago. Now more than 7 billion people live on Earth. All of us—indeed, every living organism on Earth—are the products of 3.5 billion years of evolution. We survived five major die-offs and countless smaller ones, in addition to the daily battles to remain alive. Somehow, we were always the smartest, quickest, strongest, or maybe just luckiest. Certainly we were lucky. Even species that were successful for millions of years could not cope with certain dramatic changes in the environment. The dinosaurs ruled the Earth for more than 130 million years, but they were wiped out by a meteor strike or a massive volcanic eruption or both over a relatively short time. Our hominid predecessors cheated death, at least until they reproduced. In that way, the next generations had their chance to roll the dice of survival.

Life could have easily taken another course. As Stephen Jay Gould said, we could run the clock again, and it might turn out to be another group of hominids that got lucky. He noted, "Life is a copiously branching bush, continually pruned by the grim reaper of extinction, not a ladder of predictable progress" (1989). The disappearance of all of those species eliminated those options and turned evolution in different directions. The end of the dinosaurs allowed mammals and later humans to further evolve. Even among hominids, some species died out. The Neandertals, Denisovans, and several other groups were eclipsed by their cousins, modern humans. We all carry some of their genes within us. A few prominent biologists, such as Simon Conway Morris, have argued that some form of human-like beings was inevitable (Conway Morris 2003). We just happened to be the particular combination that made it.

For 3.5 billion years, species came and went. Estimates are that more than 4 billion species have existed on the Earth at some point, and 99 percent have gone extinct (Barnosky et al. 2011). But life survived. More amazingly, in that time, life even learned how to harness death to help with reproduction and development and to defend against diseases. Programmed cell death is used to spread seeds, allow fruit to drop, extract nutrients from leaves before they change colors in the fall, and countless other actions. Importantly, it also helps to ensure that our immune system attacks only invaders and not our own tissues.

For better or worse, humans have become the dominant species on Earth. We have developed the mental capacity to change the Earth and our own fate. Now

The Biology of Death. Gary C. Howard, Oxford University Press. © Oxford University Press 2021.
DOI: 10.1093/oso/9780190687724.003.0016

Figure 16.1   Fern growing in a lava field near Kilauea.
Photograph by author.

we are in a position to defy death. In many ways, we have done well over the last hundred years or so. Through public sanitation and the miracles of modern medicine, we have learned to postpone death, and the average life span has gone from thirty or forty years to seventy or eighty years. We will likely do better in the future. There is no reason to believe that we cannot extend the average human life span to 100 years and more.

On other scores, we have not done so well. Throughout history, many species have gone extinct, and at least five times mass extinctions extinguished many forms of life on Earth. The exact cause of those events is not known, but they involved meteor strikes, volcanoes, climate change, sea level rise, and other natural causes, alone or in some combinations. Many scientists believe we are now beginning to enter another great die-off, and this one is happening much faster than the previous ones. Yet this one is not caused by natural events. Human actions are behind it. Many species are going extinct due to habitat loss, unsustainable use of natural resources, deforestation, invasive species disrupting native habitats, pollution, and—by far the most important—climate change.

Ironically, we might be one of the species that will be lost in the sixth mass extinction. Our folly is causing the global climate crisis that might put an end to us, but not likely to all life. Our predecessors slipped through the first five great die-offs. However, this is the first such global threat that we humans have faced.

It is hard to know if we will stop the threat or find some way to live with it. We might just be another of those passing dominant species that come and go. Death would then have the last laugh on us.

Whether we do or not doesn't matter to the Earth. The planet has seen this before. Dominant species come and go. Humans might just be the latest one. After each of the previous die-offs, some life was left to adapt and go on. The dinosaurs died out, and mammals rose to replace them. If humans are lost, some other group will have an opportunity to dominate the planet. Estimates are that the Earth has another 1.75 billion years before it exits the habitable zone of our solar system (Rushby et al. 2013).

Since the beginning, life has had to contend with death, but the most amazing aspect is that life evolved ways to use death. Death is more than just dying.

Glossary

Abscission. Process by which a layer of cells dies to allow leaves or fruit to drop.

Abiogenesis. The transformation of matter from nonliving to living.

Acherontic. Term that denotes the transition between life and death when macromolecules are being withdrawn from a dying leaf or other part and distributed to other living parts of the plant.

Adaptive immunity. Immune reaction that occurs after exposure to an antigen. This system is highly specific, but it takes some time for it to be activated.

Aerobic. Metabolic processes that occur in the presence of oxygen.

Amino acids. Chemical building blocks of proteins. The sequence of the amino acids determines the structure and activity of the final protein.

Anaerobic. Metabolic processes that occur in the absence of oxygen.

Antagonistic pleiotrophy theory. A single gene might affect multiple traits. Those related to reproduction are favored over long-term needs.

Antibody. Protein that binds to the specific molecular structure of another molecule in the body. Antibodies are made by cells as part of the adaptive immune system.

Antigen. Small molecules that elicit the production of antibodies by the adaptive immune system.

Apoptosis. Type of programmed cell death characterized by blebbing, cell shrinkage, nuclear fragmentation, chromatin condensation, chromosomal DNA fragmentation, and global mRNA decay.

Autoimmune disease. Condition in which a person's immune system becomes sensitized to parts of its own body.

Autophagy. Mechanism by which a cell degrades unnecessary or dysfunctional parts and recycles the material.

Bacteriophage. Virus that attacks bacteria.

Calorie restriction. Severe limitation of the number of calories. It has been associated with life span extension in a number of organisms.

Caspases. Protease enzymes that are essential in programmed cell death.

Clonal selection. Theory that states that immune cells proliferate when stimulated by a specific antigen.

Cremation. Disposal of a dead body by burning.

CRISPR. A method of editing genomes based on cutting DNA with the enzyme cas9 and a guide RNA.

Cytokines. Small signaling proteins.

Deccan Traps . Massive volcanic fields in India.

Decomposition. The breakdown of organisms after death.

Denisovans. A species or subspecies of humans that lived in Asia during the Lower and Middle Paleolithic and interbred with modern humans.

Disarticulation. Separation of bones that occurs as the tendons decay.

Disposable soma theory. Theory of aging based on the compromises that organisms must make to balance resources between reproduction and repair.

DNA damage theory. This theory suggests that aging results from an accumulation of unrepaired damaged genes.

Embalming. Infusion of preservatives into a dead body to halt or delay decay.

Enzyme. Protein that catalyzes a biochemical reaction.

Epigenetics. Heritable changes that do not involve the DNA sequence (e.g., DNA methylation, histone compositions).

Eukaryote. Cells that have a nucleus and other organelles surrounded by a membrane and organized chromosomes.

Extracellular death factor. A five-amino acid peptide that is used by bacteria in quorum sensing.

Extracellular matrix. Provides structure and support to cells. It includes several proteins, such as collagen, fibronectin, fibrinogen, elastin, and laminin, which form a very complex scaffold that cells can adhere to.

Frailty. A lack of reserves to respond to stresses; often signals an increase in the risk of death.

Genetic bottleneck. Reduction in a population due to natural disasters or human activities.

hachimoji DNA. A human-made DNA that includes up to four new nucleotides (for a total of eight) that are similar to the original purines and pyrimidines and can be used for hydrogen bonding.

Health span. The length of time in which a person is healthy.

HeLa cells. The first human cell line. The cells were taken from a dying African American woman many years ago, without her consent.

Hypersensitive response. Response by plants to an infection in which necrotic lesions to form around the infected cells and so limit the infection.

Inflamm-aging. Theory of aging that states that chronic, low-grade inflammation is a cause of aging.

iPS cells. Induced pluripotent stem cells are adult cells that have been genetically reprogrammed to become a type of stem cell.

Mass extinction. A sudden reduction in the number of species and organisms living on Earth. While extinctions happen continually, these events involve the loss of the great majority of living organisms on Earth.

Meristem. The growing part of a plant.

Metabolic syndrome. A collection of disorders (i.e., high blood pressure, high blood sugar, high cholesterol levels, and excess abdominal body fat) that increase the risk of heart disease, stroke, and diabetes.

Metamorphosis. Process in which an animal changes dramatically as it grows and differentiates.

Metastasis. In some forms of cancer, cancerous cells can break off from the primary tumor and migrate to other places in the body.

MicroRNA. Short RNA sequences that are involved in regulating protein amounts in a cell.

Mitochondria. Cell organelle that is involved in the production of energy and in programmed cell death.

Mutation accumulation theory. Theory that states that the deleterious effects of genes that affect an organism early in life are selected against because they hinder reproduction.

Neandertals. Species or subspecies of humans in Eurasia that died out about 40,000 years ago. However, they did interbreed with modern humans.

Near-death experience. Event in which a person is declared dead but later revives.

Necroptosis. Type of necrosis or programmed cell death.

Necrosis. Death of cells in a relatively unorganized manner.

Neuron. Nerve cell that is responsible for the long-range communication of nerve impulses.

Pandemic. An epidemic that involves the entire world.

Panspermia. A theory that life might have originated elsewhere in the universe and been transported to Earth.

Petrification. Process in which the organic material in a plant is replaced by minerals or the spaces in the plant are filled with minerals.

Plate tectonics. Theory stating that the Earth's crust is broken up into massive pieces that migrate over the Earth's surface.

Postmitotic. Refers to cells that have exited the cell cycle.

Prions. Infectious proteins or protein structures.

Progeria. Genetic disorder in which the characteristics of aging appear at a very accelerated rate.

Programmed cell death. Highly ordered process that result in the death of a cell. The triggering event can be internal or external. Autophagy and apoptosis are two types of programmed cell death.

Prokaryote. Cells lacking a nucleus and other organelles that are surrounded by a membrane. These include bacteria and blue-green algae.

Proteostasis. The process of maintaining the correct types and levels of proteins in a cell. It involves both production and destruction of proteins.

Protista. Any eukaryote that is not an animal, plant, or fungus.

Pyroptosis. A type of programmed cell death that involves a distinct set of caspases and results in a highly inflammatory reaction.

Quorum sensing. The ability of bacteria to communicate with each other by exchanging diffusible signaling molecules to determine if they must reduce the number of individuals in the colony to ensure long-term survival of the colony.

R factor. Proteins involved in resistance against plant diseases.

Reactive oxygen species. Compounds that readily react with biological molecules; they include peroxides, superoxides, and hydroxyl radicals.

Recency. The tendency of people to weigh recent events as more important than those from the past.

Red queen. Theory in which one organism has to continually evolve new defenses to keep up with its predators.

Reliability theory. Theory of aging that suggests that humans are complex systems of many parts and multiple redundant systems.

Resveratrol. Compound associated with enhanced life span in humans.

Rigor mortis. The contraction of the muscles that occurs after death.

Secretome. The proteins and other compounds that are secreted by a cell.

Senescence. Usually refers to the loss of biological functions in cells or whole organisms.

Senescence-associated secretory phenotype. Refers to the many different compounds, including cytokines, chemokines, growth factors, and proteases, that are secreted by an aging cell.

Sirtuins. Proteins involved in maintaining protein homeostasis or proteostasis in cells. They depend on NAD as a cofactor.

Supercontinent. An assembly of all or most of the Earth's continents. These have occurred at various times in the geologic past.

Synthetic biology. Involves redesigning organisms to have new abilities.

Taphonomy. Study of the transition of animal remains from the biosphere into the lithosphere.

Terran. Related to the Earth.

Telomers. Structures at the end of chromosomes that are related to aging.

Thrifty gene hypothesis. The idea that for millennia humans lived in environments where food was scarce, and our ancestors adapted by favoring genes that enhanced fat deposition.

Transcription factor. Protein that regulates the genes.

Ubiquitin-proteasome system. Mechanism in cells in which proteins to be degraded are labeled with ubiquitin and then degraded in the proteasome.

Viroids. Particles that have a genome of a short strand of circular, single-stranded RNA but completely lack a protein coat.

Vitalism. The concept that living organisms are governed by nonphysical principles and thus are fundamentally different from nonliving material.

References

Aalto-Setälä K, Conklin BR, Lo B (2009) Obtaining consent for future research with induced pluripotent cells: Opportunities and challenges. *PLoS Biol 7*(2): e1000042.

Abdel-Maksouda G, El-Aminb AR (2011) A review on the materials used during the
mummification processes in ancient Egypt. *Mediterr Archaeol Archaeom 11*: 129–150.

Ackermann M, Chao L, Bergstron CT, Doebeli M (2007) On the evolutionary origin of
aging. *Aging Cell 6*: 235–244.

Ahnstedt H, Patrizz A, Chauhan A, Roy-O'Reilly M, Furr JW, Spychala MS, D'Aigle J, Blixt
FW, Zhu L, Bravo Alegria J, McCullough LD (2020) Sex differences in T cell immune
responses, gut permeability and outcome after ischemic stroke in aged mice. *Brain,
Behavior, and Immunity 87*: 556–567.

Albert V, Hall MN (2015) mTOR signaling in cellular and organismal energetics. *Curr
Opin Cell Biol 33*: 55–66.

Alberti S, Hyman AA (2016) Are aberrant phase transitions a driver of cellular aging?
Bioessays 38: 959–968.

Alirol E, Getaz L, Stoll B, Chappuis F, Loutan L (2010) Urbanisation and infectious diseases in a globalised world. *Lancet Infect Dis 10*: 131–141.

Alkire MT, Hudetz AG, Tononi G (2008) Consciousness and anesthesia. *Science
322*: 876–880.

Allocati N, Masulli M, Di Ilio C, De Laurenzi V (2015) Die for the community: An over-
view of programmed cell death in bacteria. *Cell Death Dis 6*: e1609.

Alvarez LW, Alvarez W, Asaro F, Michel HV (1980) Extraterrestrial cause for the
Cretaceous-Tertiary Extinction. *Science 208*: 1095–1108.

Alzheimer's Association (2016) Alzheimer's disease facts and figures. *Alzheimer's &
Dementia 12*: 459–509.

Alzheimer's Association (2019) *2019 Alzheimer's disease facts and figures.* https://www.
alz.org/alzheimers-dementia/facts-figures.

Amos W, Harwood J (1998) Factors affecting levels of genetic diversity in natural populations. *Philos Trans R Soc Lond B Biol Sci 353*: 177–186.

Andersen KG, Rambaut A, Lipkin WI, Holmes EC, Garry RF (2020) The proximal origin
of SARS-CoV-2. *Nat Med 26*: 450–452.

Anderson JR (2016) Comparative thanatology. *Curr Biol 26*: R543–R576.

Anderson JR (2018) Chimpanzees and death. *Phil Trans R Soc B 373*: 20170257.

Arandjelovic S, Ravichandran KS (2016) Phagocytosis of apoptotic cells in homeostasis.
Nat Immunol 26: 907–917.

Aristotle (1910) *The History of Animals.* Translated by D'Arcy Wentworth Thompson.
Book IV of the Works of Aristotle. Oxford: Clarendon Press. http://classics.mit.edu/
Aristotle/history_anim.8.viii.html.

Ashford TP, Porter KR (1962) Cytoplasmic components in hepatic cell lysosomes. *J Cell
Biol 12*: 198–202.

ASPS (2017) *American Society of Plastic Surgeons releases report showing national average
of cosmetic surgical fees in 2016.* American Society of Plastic Surgeons. https://www.

plasticsurgery.org/news/press-releases/more-than-16-billion-spent-on-cosmetic-plastic-surgery. Retrieved on January 18, 2021.

ASPS (2018) *2018 Plastic Surgery Statistics Report*. American Society of Plastic Surgeons. https://www.plasticsurgery.org/documents/News/Statistics/2018/plastic-surgery-statistics-full-report-2018.pdf. Retrieved on January 18, 2021.

Astrobiology staff (2003) Life's working definition: Does it work? *Astrobiology Magazine*. https://www.astrobio.net/origin-and-evolution-of-life/lifes-working-definition/. Retrieved on June 19, 2021.

Baehrecke EH (2003) Autophagic programmed cell death in *Drosophila*. *Cell Death Differ* 10: 940–945.

Barbi E, Lagona F, Marsili M, Vaupel JW, Wachter KW (2018) The plateau of human mortality: Demography of longevity pioneers. *Science 360*: 1459–1461.

Barlow AD, Nicholson ML, Herbert TP (2013) Evidence for rapamycin toxicity in pancreatic β-cells and a review of the underlying molecular mechanisms. *Diabetes* 62: 2674–2682.

Barnosky AD, Matzke N, Tomiya S, Wogan GOU, Swartz B, Quental TB, Marshall C, McGuire JL, Lindsey EL, Maguire KC, Mersey B, Ferrer EA (2011) Has the Earth's sixth mass extinction already arrived? *Nature 471*: 51–57.

Basisty N, Kale A, Jeon OH, Kuehnemann C, Payne T, Rao C, Holtz A, Shah S, Sharma V, Ferrucci L, Campisi J, Schilling B (2020) A proteomic atlas of senescence-associated secretomes for aging biomarker development. *PLoS Biol 18*(1): e3000599.

Bayles KW (2007) The biological role of death and lysis in biofilm development. *Nat Rev Microbiol 5*: 721–726.

Bayles KW (2014) Bacterial programmed cell death: Making sense of a paradox. *Nat Rev Microbiol 12*: 63–69.

Becker E (1973) *Denial of Death*. Free Press, New York.

Beecher HK, Adams RD, Barger AC, Curran WJ, Denny-Brown D, Farnsworth DL, Folch-Pi J, Mendelsohn EI, Merrill JP, Murray J, Potter R, Schwab R, Sweet W (1968) A definition of irreversible coma. *JAMA 205*: 85–88.

Behrents RG (2016) Lucy fell from a tree and plunged 40 feet to her death. *Am J Orthod Dentofacial Orthop 150*: 719–722.

Belsky DW, Caspi A, Houts R, Cohen HJ, Corcoran DL, Danes A, Harrington HL, Israel S, Levine ME, Schaefer JD, Sugden K, Williams B, Yashin AI, Poulton R, Moffitt TE (2015) Quantification of biological aging in young adults. *Proc Natl Acad Sci USA 112*: E4104–E4110.

Beltrán-Sánchez H, Austad SN, Finch CE (2018) Comment on "The plateau of human mortality: Demography of longevity pioneers." *Science 361*: eaav1200.

Benner SA (2010) Defining life. *Astrobiology 10*: 1021–1030.

Berghe TV, Vanlangenakker N, Parthoens E, Deckers W, Devos M, Festjens N, Guerin CJ, Brunk UT, Declercq W, Vandenabeele P (2010) Necroptosis, necrosis and secondary necrosis converge on similar cellular disintegration features. *Cell Death Differ17*: 922–930.

Bernat JL (2013) Controversies in defining and determining death in critical care. *Nat Rev Neurology 9*: 164–173.

Bhattacharjee A, Sen MR, Prakash P, Anupurba S (2008) Role of β-lactamase inhibitors in enterobacterial isolates producing extended-spectrum β-lactamases. *J Antimicrob Chemother 61*: 309–314.

Bianconi E, Piovesan A, Facchin F, Beraudi A, Casadei R, Frabetti F, Vitale L, Pelleri MC, Tassani S, Piva F, Perez-Amodio S, Strippoli P, Canaider S (2013) An estimation of the number of cells in the human body. *Ann Hum Biol 40*: 463–471.

Bierce A (1911) *The Devil's Dictionary.* Project Gutenberg. https://www.gutenberg.org/files/972/972-h/972-h.htm. Retrieved January 18, 2021.

Bijma J, Portner HO, Yesson C, Rogers AD (2013) Climate change and the oceans—What does the future hold? *Mar Poll Bull 74*: 405–505.

Binder M, Roberts C, Spencer N, Antoine D, Cartwright C (2014) On the antiquity of cancer: Evidence for metastatic carcinoma in a young man from ancient Nubia (c. 1200BC). *PLOS ONE 9*(3): e90924.

Blackburn EH, Gall JG (1978) A tandemly repeated sequence at the termini of the extra-chromosomal ribosomal RNA genes in Tetrahymena. *J Mol Biol 120*: 33–53.

Blackburn TJ, Olsen PE, Bowring SA, McLean NM, Kent DV, Puffer J, McHone G, Rasbury ET, El-Touhami M (2013) Zircon U-Pb geochronology links the end-Triassic extinction with the Central Atlantic Magmatic Province. *Science 340*: 941–945.

Blagosklonny MV (2010) Why the disposable soma theory cannot explain why women live longer and why we age. *Aging (Albany NY) 2*: 884–887.

Blanke O, Arzy S (2005) The out-of-body experience: Disturbed self-processing at the temporo-parietal junction. *The Neuroscientist 1*: 16–24.

Boers SN, van Delden JM, Bredenoord AL (2019) Organoids as hybrids: Ethical implications for the exchange of human tissues. *J Med Ethics 45*: 131–139.

Bojesen SE (2013) Telomeres and human health. *Journal of Internal Medicine 274*: 399–413.

Bókkon I, Mallick BN, Tuszynski JA (2013) Near death experiences: A multidisciplinary hypothesis. *Front Hum Neurosci 7*: 533.

Bonani G, Ivy SD, Hajdas I, Niklaus T, Suter M (1994) AMS 14C age determinations of tissue, bone, and grass samples from the Otztal ice man. *Radiocarbon 36*: 247–250.

Bordone L, Guarente L (2005) Calorie restriction, SIRT1 and metabolism: Understanding longevity. *Nat Rev Mol Cell Biol 6*: 298–305.

Boskey AL, Coleman R (2010) Aging and bone. *J Dent Res 89*: 1333–1348.

Bowles S (2009) Did warfare among ancestral hunter-gatherers affect the evolution of human social behaviors? *Science 324*: 1293–1298.

Boyce JM, Shone GR (2006) Effects of ageing on smell and taste. *Postgrad Med J 82*: 239–241.

Brill A, Torchinsky A, Carp H, Toder V (1999) The role of apoptosis in normal and ab-normal embryonic development. *J Assist Reprod Genet 16*: 512–519.

Broad JB, Gott M, Kim H, Boyd M, Chen H, Connolly MJ (2013) Where do people die? An international comparison of the percentage of deaths occurring in hospital and residential aged care settings in 45 populations, using published and available statistics. *Int J Public Health 58*: 257–267.

Brokowski C, Adli M (2018) CRISPR ethics: Moral considerations for applications of a powerful tool. *J Mol Biol 431*: 88–101.

Brunham LR, Hayden MR (2013) Hunting human disease genes: Lessons from the past, challenges for the future. *Hum Genet 132*: 603–617.

Buck LT, Stringer CB (2014) Homo heidelbergensis. *Curr Biol 24*: R214–215.

Bulterijs S, Hull RS, Bjork VCE, Roy AG (2015) It is time to classify biological aging as a disease. *Front Genet 6*: 205.

Burgess SD, Bowring S, Shen SZ (2014) High-precision timeline for Earth's most severe extinction. *Proc Natl Acad Sci USA 111*: 3316–3321.

Burian A, de Reuille PB, Kuhlemeier C (2016) Patterns of stem cell divisions contribute to plant longevity. *Curr Biol 26*: 1385–1394.

Burke BL, Martens A, Faucher EH (2010) Two decades of terror management theory: A meta-analysis of mortality salience research. *Personality and Social Psychology Review 14*: 155–195.

Buts L, Lah J, Dao-Thi MH, Wyns, Loris R (2005) Toxin-antitoxin modules as bacterial metabolic stress managers. *Trends Biochem Sci 30*: 672–679.

Buzzi G (2002) Near death experiences. *The Lancet 359*: P2116–P2117.

Cairns J (1978) *Cancer. Science and Society* W. H. Freeman and Company, San Francisco.

Campbell-Thompson M, Fu A, Kaddis JS, Wasserfall C, Schatz DA, Pugliese A, Atkinson MA (2016) Insulitis and b-Cell Mass in the Natural History of Type 1 Diabetes. *Diabetes 65*: 719–731.

Cano RJ, Borucki MK (1995) Revival and identification of bacterial spores in 25- to 40-million-year-old Dominican amber. *Science 268*: 1060–1064.

Cao H, Hegele RA (2003) *LMNA* is mutated in Hutchinson-Gilford progeria (MIM 176670) but not in Wiedemann-Rautenstrauch progeroid syndrome (MIM 264090). *Journal of Human Genetics 48*: 271–274.

Capasso LL (2004) Antiquity of cancer. *Int J Cancer 113*: 2–13.

Caplan A (2015) How can aging be thought of as anything other than a disease. In *Handbook of the Philosophy of Medicine* (Eds: T. Schramme, S Edwards) Springer. Dordrecht, The Netherlands.

Carnes BA, Olshansky SJ, Grahn D (2003) Biological evidence for limits to the duration of life. *Biogerontology 4*: 31045.

Case A, Deaton A (2015) Rising morbidity and mortality in midlife among white non-Hispanic Americans in the 21st century. *Proc Natl Acad Sci 112*: 15078–15083.

CDC (2020a) *National Diabetes Statistics Report, 2020*. Centers for Disease Control and Prevention. https://www.cdc.gov/diabetes/data/statistics-report/index.html.

CDC (2020b) *CDC COVID-19 Data Tracker*. Centers for Disease Control and Prevention. https://www.cdc.gov/coronavirus/2019-ncov/index.html.

CDC (2020c) *Leading Causes of Death, 1900–1998*, Centers for Disease Control and Prevention. retrieved October 1, 2020. https://www.cdc.gov/nchs/data/dvs/lead1900_98.pdf.

Cech TR (1986) RNA as an enzyme. *Sci. Amer.* 11: 76–84.

Chan SRWL, Blackburn EH (2004) Telomeres and telomerase. *Phil. Trans. R. Soc. Lond. B 359*: 109–121.

Charlesworth B (2000) Fisher, Medawar, Hamilton and the evolution of aging. *Genetics 156*: 927.

Charlesworth B (2001) Patterns of age-specific means and genetic variances of mortality rates predicted by the mutation-accumulation theory of ageing. *J Theor Biol 210*: 47–65.

Chauhan A, Moser H, McCullough LD (2017) Sex differences in ischaemic stroke: Potential cellular mechanisms. *Clin Sci (Lond) 131*: 533–552.

Cheng X, Ferrell Jr. JE (2018) Apoptosis propagates through the cytoplasm as trigger waves. *Science 361*: 607–612.

Chetty R, Stepner M, Abraham S, Lin S, Scuderi B, Turner N, Bergeron A, Cutler D (2016) The association between income and life expectancy in the United States, 2001–2014. *JAMA 315*: 1750–1766.

Chiao YA, Rabinovitch PS (2015) The aging heart. *Cold Spring Harb Perspect Med* 5: a025148.

Chibani-Chennoufi S, Bruttin A, Dillmann ML, Brussow H (2004) Phage–host interaction: An ecological perspective. *J Bacteriol 186*: 3677–3686.

Choi BCK, Hunter DJ, Tsou W, Sainsbury P (2005) Diseases of comfort: Primary cause of death in the 22nd century. *J Epidemial Community Health 59*: 1030–1034.

Choi BCK, McQueen DV, Puska P, Douglas KA, Ackland M, Campostrini S, Barcelo A, Stachenko S, Mokdad AH, Granero R, Corber SJ, Valleron AJ, Skinner HA, Potemkina R, Lindner MC, Zakus D, de Salazar LM, Pak AWP, Ansari Z, Zevallos JC, Gonzalez M, Flahault A, Torres RE (2008) Enhancing global capacity in the surveillance, prevention, and control of chronic diseases: seven themes to consider and build upon. *J Epidemiol Community Health 62*: 391–397.

Clark SL Jr (1957) Cellular differentiation in the kidneys of newborn mice studies with the electron microscope. *J Biophys Biochem Cytol 3*: 349–362.

Clegg RJ, Dyson RJ, Kreft JU (2014) Repair rather than segregation of damage is the optimal unicellular aging strategy. *BMC Biology 12*: 52.

Cleland CE, Chyba CF (2002) Defining life. *Origins of Life and Evolution of the Biosphere 32*: 387–393.

Collins FS, McKusick VA (2001) Implications for the Human Genome Project for medical science. *JAMA 285*: 540–544.

Colson P, de Lamballerie X, Fournous G, Raoult D (2012) Reclassification of giant viruses composing a fourth domain of life in the new order Megavirales, *Intervirology 55*: 321–332.

Colson P, La Scola B, Levasseur A, Caetano-Anollés G, Raoult D (2020) Mimivirus: leading the way in the discovery of giant viruses of amoebae. *Nat Rev Microbiol 15: 243–254.*

Connolly RC (1985) Lindow man: Britain's prehistoric bog body. *Anthropology Today 1*: 15–17.

Conradt B, Wu YC, Xue D (2016) Programmed cell death during *Caenorhabditis elegans* development. *Genetics 203*: 1533–1562.

Conway MS (2003) *Life Is Solution: Inevitable Humans in a Lonely Universe.* Cambridge: Cambridge University Press.

Cook MR (2005) *The Faces of Science.* W. W. Norton, New York.

Cornforth DM, Popat R, McNally L, Gurney J, Scott-Phillips TC, Ivens A, Diggle SP, Brown SP (2014) Combinatorial quorum sensing allows bacteria to resolve their social and physical environment. *Proc Natl Acad Sci USA 111*: 4280–4284.

Costanzo A, Fausti F, Spallone G, Moretti F, Narcisi A, Botti E (2015) Programmed cell death in the skin. *Int J Dev Biol 59*: 73–78.

Cruz Reyes J, Tata JR (1995) Cloning, characterization and expression of two *Xenopus* bcl-2-like cell-survival genes. *Gene 158*: 171–179.

Cummins N, Kelly M, Grada CO (2015) Living standards and plague in London, 1560–1665. *The Economic History Review 69*: 3–34.

Curtis JR, Vincent JL (2010) Ethics and end-of-life care for adults in the intensive care unit. *Lancet 376*: P1347–1353.

Czabotar PE, Lessene G, Strasser A, Adams JM (2014) Control of apoptosis by the BCL-2 protein family: Implications for physiology and therapy. *Nat Rev Mol Cell Biol 15*: 49–63.

Dagdas YF, Belhaj K, Maqbool A, Chaparro-Garcia A, Pandey P, Petre B, Tabassum N, Cruz-Mireles N, Hughes RK, Sklenar J, Win J, Menke F, Findlay K, Banfield MJ,

Kamoun S, Bozkurt TO (2016) An effector of the Irish potato famine pathogen antag-
onizes a host autophagy cargo receptor. *eLife* 2016;5: e10856.

Dai DF, Chiao YA, Marcinek DJ, Szeto HH, Rabinovitch PS (2014) Mitochondrial oxidative stress in aging and healthspan. *Longev Healthspan* 3: 6.

Dangl JL, Jones JDG (2001) Plant pathogens and integrated defence responses to infection. *Nature 411*: 826–833.

D'Aquila P, Montesanto A, De Rango F, Guarasci F, Passarino G, Bellizzi D (2019)
Epigenetic signature: Implications for mitochondrial quality control in human aging.
Aging (Albany) 11: 1240–1251.

Davis A, McMahon CM, Pichora-Fuller KM, Russ S, Lin F, Olusanya BO, Chadha S,
Tremblay KL (2016) Aging and hearing health: The life-course approach. *Gerontologist*
56: S256–S267.

de Beer J, Bardoutsos A, Janssen F (2017) Maximum human lifespan may increase to
125 years. *Nature 546*: E16–E17.

de Boer W (2017) Upscaling of fungal-bacterial interactions: From the lab to the field.
Curr Opin Microbiol 37: 35–41.

De Duve C, Wattiaux R (1966) Functions of lysosomes. *Annu Rev Physiol 28*: 435–492.

Deelen J, Kettunen J, Fischer K, van der Spek A, Trompet S, Kastenmüller G, Boyd A,
Zierer J, van den Akker EB, Ala-Korpela M, Amin N, Demirkan A, Ghanbari M, van
Heemst D, Ikram MA, van Klinken JB, Mooijaart SP, Peters A, Salomaa V, Sattar N,
Spector TD, Tiemeier H, Verhoeven A, Waldenberger M, Würtz P, Smith GD, Metspalu
A, Perola M, Menni C, Geleijnse JM, Drenos F, Beekman M, Jukema JW, van Duijn
CM, Slagboom PE (2019) A metabolic profile of all-cause mortality risk identified in an
observational study of 44,168 individuals. *Nat Communications 10*: 3346.

Dickman M, Williams B, Li Y, Figueiredo P, Wolpert T (2017) Reassessing apoptosis in
plants. *Nat Plants 3*: 773–779.

Dickman MB, de Figueiredo P (2013) Death be not proud—Cell death control in plant
fungal interactions. *PLoS Pathog 9*(9): e1003542.

Dickstein DL, Kabaso D, Rocher AB, Luebke JI, Wearne SL, Hof PR (2007) Changes in the
structural complexity of the aged brain. *Aging Cell 6*: 275–284.

Diehr P, Williamson J, Burke GL, Psaty BM (2002) The aging and dying processes and the
health of older adults. *J Clin Epidemiol 55L*: 269–278.

Dirks PHGM, Berger LR, Roberts EM, Kramers JD, Hawks J, Randolph-Quinney PS,
Elliott M, Musiba CM, Churchill SE, de Ruiter DJ, Schmid P, Backwell LR, Belyanin
GA, Boshoff P, Hunter KL, Feuerriegel EM, Gurtov A, Harrison J, Hunter R, Kruger
A, Morris H, Makhubela TV, Peixotto B, Tucker S (2015) Geological and taphonomic
context for the new hominin species *Homo naledi* from the Dinaledi Chamber, South
Africa. *eLife 4*: e09561.

Dirks PHGM, Roberts EM, Hilbert-Wolf H, Kramers JD, Hawks J, Dosseto A, Duval M,
Elliott M, Evans M, Grun R, Hellstrom J, Herries AIR, Joannes-Boyau R, Makhubela
TV, Placzek CJ, Robbins J, Spandler C, Wiersma J, Woodhead J, Berger LR (2017) The
age of *Homo naledi* and associated sediments in the Rising Star Cave, South Africa.
eLife 6: e24231.

Distéfano AM, Martin MV, Córdoba JP, Bellido AM, D'Ippólito S, Colman SL, Roldán JA,
Bartoli CG, Zabeleta EJ, Fiol DF, Stockwell BR, Dixon SJ, Pagnussat GC (2017) Heat
stress induces ferroptosis-like cell death in plants. *J Cell Biol 216*: 463–476.

Doitsh G, Galloway NL, Geng X, Yang Z, Monroe KM, Zepeda O, Hunt PW, Hatano H,
Sowinski S, Muñoz-Arias I, Greene WC (2014) Cell death by pyroptosis drives CD4 T-
cell depletion in HIV-1 infection *Nature 505*: 509–514.

Dong X, Milholland B, Vijg J (2016) Evidence for a limit to human lifespan. *Nature* 538: 257–259.

Dowling-Lacey D, Mayer JF, Jones E, Bocca S, Stadtmauer L, Oehniger S (2011) Live birth from a frozen–thawed pronuclear stage embryo almost 20 years after its cryopreservation. *Fertility and Sterility 95*: 1120.e1–1120.e3.

Dreier JP, Major S, Foreman B, Winkler MKL, Kang EJ, Milakara D, Lemale CL, DiNapoli V, Hinzman JM, Woitzik J, Andaluz N, Carlson A, Hartings JA (2018) *Ann Neurol 83*: 295–310.

Efremov IA (1940) Taphonomy: A new branch of paleontology. *Pan-American Geology 74*: 81–93.

Erikson GA, Bodian KL, Rueda M, Molparia B, Scott ER, Scott-Van Zeeland AA, Topol SE, Wineinger NE, Niederhuber JE, Topol EJ, Torkamani A (2016) Whole-genome sequencing of a healthy aging cohort. *Cell 165*: 1002–1011.

Erives AJ (2017) Phylogenetic analysis of the core histone doublet and DNA topo II genes of Marseilleviridae: evidence of proto-eukaryotic provenance. *Epigenetics & Chromatin*, 2017; 10: 55.

Erwin DH (2014) Temporal acuity and the rate and dynamics of mass extinctions. *Proc Natl Acad Sci USA 111*: 3203–3204.

Exbrayat JM, Moudilou EN, Abrouk L, Brun C (2012) Apoptosis in amphibian development. *Advances in Bioscience and Biotechnology 3*: 669–678.

Ezzati M, Vander Hoorn S, Lawes CMM, Leach R, James WPT, Lopez AD, Rodgers A, Murray CJL (2005) Rethinking the "diseases of affluence" paradigm: Global patterns of nutritional risks in relation to economic development. *PLoS Med 3*: e133.

Farahany NA, Greely HT, Giattino CM (2019) Part-revived pig brains raise ethical questions. *Nature 568*: 299–302.

Fauci AS, Morens DM (2016) Zika Virus in the Americas—Yet another arbovirus threat. *N Engl J Med 374*: 601–604. DOI: 10.1056/NEJMp1600297.

Ferreira PG, Muñoz-Aguirre M, Reverter F, Sá Godinho CP, Sousa A, Amadoz A, Sodaei R, Hidalgo MR, Pervouchine D, Carbonell-Caballero J, Nurtdinov R, Breschi A, Amador R, Oliveira P, Cubuk C, Çurado J, Aguet F, Oliveira C, Dopazo J, Sammeth M, Ardlie KG, Guigó R (2018) The effects of death and post-mortem cold ischemia on human tissue transcriptome. *Nat Commu 9*: 490.

Figge MT, Reichert AS, Meyer-Hermann M, Osiewacz HD (2012) Deceleration of fusion-fission cycles improves mitochondrial quality control during aging. *PLoS Comput Biol 8*, e1002576.

Finerana PC, Blowera TR, Fouldsa IJ, Humphrey DP, Lilleya KS, Salmond GPC (2009) The phage abortive infection system, ToxIN, functions as a protein–RNA toxin-antitoxin pair. *Proc Natl Acad Sci USA 106*: 894–899.

Finkbeiner S (2020) The autophagy lysosomal pathway and neurodegeneration. *Cold Spring Harb Perspect Biol 12*(3): a033993.

Finnegan S, Heim NA, Peters SE, Fischer WW (2012) Climate change and the selective signature of the Late Ordovician mass extinction. *Proc Natl Acad Sci USA 109*: 6829–6834.

Fisher CB, Layman DM (2018) Genomics, big data, and broad consent: A new ethics frontier for prevention science. *Prev Sci 19*: 871–879.

Föller M, Huber SM, Lang F (2008) Erythrocyte programmed cell death. *Life 60*: 661–668.

Forterre P (2010) Defining life: The virus viewpoint. *Orig Life Evol Biosph 40*: 151–160.

Fortney K, Dobriban E, Garagnani P, Pirazzini C, Monti D, Mari D, Atzmon G, Barzilai N, Franceschi C, Own AB, Kim SK (2015) Genome-wide scan informed by age-related disease identifies loci for exceptional human longevity. *PLoS Genetics 11*(12): e1005728.

Franceschi C, Bonafè M, Valensin S, Olivieri F, De Luca M, Ottaviani E, de Benedictus G (2000) Inflamm-aging. An evolutionary perspective on immunosenescence. *Ann NY Acad Sci 908*: 244–254.

Franceschi C, Garagnani P, Morsiani C, Conte M, Santoro A, Grignolio A, Monti D, Capri M, Salvioli S (2018) The continuum of aging and age-related diseases: Common mechanisms but different rates. *Front Med.* https://doi.org/10.3389/fmed.2018.00061.

Franceschi C, Garagnani P, Vitale G, Capri M, Salvioli S (2017) Inflammaging and 'Garbaging'. *Trends Endocrinol Metab 28*: 199–212.

Francis LP (2014) Genomic knowledge sharing: A review of the ethical and legal issues. *Appl Transl Genom 3*: 111–115.

Freitas AA, de Magalhães JP (2011) A review and appraisal of the DNA damage theory of ageing. *Mutat Res 728*: 12–22.

Freitas RA (2002) *Death is an outrage.* Lecture presented at the Fifth Alcor Conference on Extreme Life Extension, November 16, Newport Beach, CA. www.rfreitas.com/Nano/DeathIsAnOutrage.htm.

Freund AM, Nikitin FJ, Ritter JO (2009) Psychological consequences of longevity the increasing importance of self-regulation in old age. *Hum Devel 52*: 1–37.

Fricker M, Tolkovsky AM, Borutaite V, Coleman M, Brown GC (2018) Neuronal cell death. *Physiol Rev 98*: 813–880.

Fu Q, Hajdinjak M, Moldovan OT, Constantin S, Mallick S, Skoglund P, Patterson N, Rohland N, Lazaridis I, Nickel B, Viola B, Prüfer K, Meyer M, Kelso J, Reich D, Pääbo S (2015) An early modern human from Romania with a recent Neanderthal ancestor. *Nature 524*: 216–219.

Fuchs Y, Steller H (2011) Programmed cell death in animal development and disease. *Cell 147*: P742–758.

Galluzzi L, Vitale I, Aaronson S et al. (2018) Molecular mechanisms of cell death: Recommendations of the Nomenclature Committee on Cell Death 2018. *Cell Death Differ 25*: 486–541.

Gao GF (2018) From "A"IV to "Z"IKV: Attacks from emerging and re-emerging pathogens. *Cell 172*: 1157–1159.

Gat A (2015) Proving communal warfare among hunter-gatherers: The quasi-Rousseauan error. *Evolutionary Anthropology 24*: 111–126.

Gavrilov LA, Gavrilova NS (2003) The quest for a general theory of aging and longevity. *Sci. Aging Knowledge Environ 2003*(28): RE5.

Gelens L, Anderson GA, Ferrell JE Jr (2014) Spatial trigger waves: Positive feedback gets you a long way. *Mol Biol Cell 25*: 3486–3493.

Gems D (2011) Tragedy and delight: The ethics of decelerated ageing. *Phil Trans R Soc B 366*: 108–112.

Gems D (2015) The aging-disease false dichotomy: Understanding senescence as pathology. *Front Genet 6*: 212.

Gibbons A (1995) The mystery of humanity's missing mutations. *Science 267*: 35–36.

Gibson DG, Glass JI, Lartigue C, Noskov VN, Chuang RY, Algire MA, Benders GZ, Montague MG, Ma L, Moodie MM, Merryman C, Vashee S, Krishnakumar R, Assad-Garcia N, Andrews-Pfannkoch C, Denisova EA, Young L, Qi ZQ, Segall-Shapiro TH, Calvey CH, Parmar PP, Hutchison III CA, Smith HO, Venter JC (2010) Creation of a bacterial cell controlled by a chemically synthesized genome. *Science 329*: 52–56.

Gierman HJ, Fortney K, Roach JC, Coles NS, Li H, Glusman G, Markov GJ, Smith JD, Hood L, Coles LS, Kim SK (2014) Whole-genome sequencing of the world's oldest people. *PLoS ONE 9*: e112430.

Gilbert WH (2018) Hominins, Early. *The International Encyclopedia of Anthropology.* https://doi.org/10.1002/9781118924396.wbiea1920. Retrieved on January 18, 2021.

Gillespie R (1997) Burnt and unburnt carbon: Dating charcoal and burnt bone from the Willandra Lakes, Australia. *Radiocarbon 39*: 239–250.

Gilmore AP (2005) Anoikis. *Cell Death Differ12*: 1473–1477.

Glass NL, Dementhon K (2006) Non-self recognition and programmed cell death in filamentous fungi. *Curr Opin Microbiol 9*: 553–558.

Glücksmann A (1951) Cell deaths in normal vertebrate ontogeny. *Biol Rev 29*: 59–86.

Go AS, Mozaffarian D, Roger VL, Benjamin EJ, Berry JD, Blaha MJ, Dai S, Ford ES, Fox CS, Franco S, Fullerton HJ, Gillespie C, Hailpern SM, Heit JA, Howard VJ, Huffman MD, Judd SE, Kissela BM, Kittner SJ, Lackland DT, Lichtman JH, Lisabeth LD, Mackey RH, Magid DJ, Marcus GM, Marelli A, Matchar DB, McGuire DK, Mohler ER 3rd, Moy CS, Mussolino ME, Neumar RW, Nichol G, Pandey DK, Paynter NP, Reeves MJ, Sorlie PD, Stein J, Towfighi A, Turan TN, Virani SS, Wong ND, Woo D, Turner MB, American Heart Association Statistics Committee and Stroke Statistics Subcommittee (2014) Heart disease and stroke statistics--2014 update: a report from the American Heart Association. *Circulation 129*: e28–e292.

Gonçalves AP, Heller J, Daskalov A, Videira A, Glass NL (2017) Regulated forms of cell death in fungi. *Front Microbiol 8*: 1837.

Gong YN, Crawford JC, Heckmann BL, Green DR (2019) To the edge of cell death and back. *FEBS J 286*: 430–440.

Gong YN, Guy C, Olauson H, Becker JU, Yang M, Fitzgerald P, Linkermann A, Green DR (2017) ESCRT-III acts downstream of MLKL to regulate necroptotic cell death and its consequences. *Cell 169*: 286–300.

Gould SJ (1989) *Wonderful Life. The Burgess Shale and the Nature of History.* WW Norton, New York. p. 51.

Goya RG, Lehmann M, Chiavellini P, Chanatelli-Mallat M, Hereñú CB, Brown OA (2018) Rejuvenation by cell reprogramming: a new horizon in gerontology. *Stem Cell Research & Therapy 9*: 349.

Green DR (2005) Apoptotic pathways: Ten minutes to dead. *Cell 121*: 671–674.

Greider CW, Blackburn EF (1985) Identification of a specific telomere terminal transferase activity in Tetrahymena extracts. *Cell 43*: 405–413.

Greyson B (1983) The near-death experience scale: Construction, reliability, and validity. *Journal of Nervous and Mental Disease 171*: 369–375.

Guarente L (2007) Sirtuins in aging and disease. *Cold Spring Harb Symp Quant Biol 72*: 483–488.

Gunawardena AH (2008) Programmed cell death and tissue remodeling in plants. *J Exp Bot 59*: 445–451.

Hajat S, Vardoulakis S, Heaviside C, Eggen B (2014) Climate change effects on human health: projections of temperature-related mortality for the UK during the 2020s, 2050s and 2080s. *Journal of Epidemiology and Community Health 0*: 1–8.

Hamann JC, Surcel A, Chen R, Teragawa C, Albeck JG, Robinson DN, Overholtzer M (2017) Entosis is induced by glucose starvation. *Cell Rep 20*: 201–210.

Hammer MF, Woerner AE, Mendez FL, Watkins JC, Wall JD (2011) Genetic evidence for archaic admixture in Africa. *Proc Natl Acad Sci USA 108*: 15123–15128.

Han GZ (2019) Origin and evolution of the plant immune system. *New Phytologist 222*: 70–83.

Hanamata S, Sawada J, Ono S, Ogawa K, Fukunaga T, Nonomura K-I, Kimura S, Kurusu T, Kuchitsu K (2020) Impact of autophagy on gene expression and tapetal programmed cell death during pollen development in rice. Front. Plant Sci 11:172. https://doi.org/ 10.3389/fpls.2020.00172.

Hansen M, Kennedy BK (2016) Does longer lifespan mean longer healthspan? *Trends in Cell Biology 26*: 565–568.

Hansen M, Taubert S, Crawford D, Libina N, Lee SJ, Kenyon C (2007) Lifespan extension by conditions that inhibit translation in *Caenorhabditis elegans. Aging Cell 6*: 95–110.

Hanto DW, Veatch RM (2010) Uncontrolled donation after circulatory determination of death (UCCDD) and the definition of death. *Am J Transplant 11*: 1351–1352.

Happo L, Strasser A, Cory S (2012) BH3-only proteins in apoptosis at a glance. *Journal of Cell Science 125*: 1081–1087.

Harada CN, Natelson Love MC, Triebel K (2013) Normal cognitive aging. *Clin Geriatr Med 29*: 737–752.

Harbut RF (2019) How should one live everlasting life? *AMA J Ethics 21*: E470–474.

Harris J (2005) Scientific research is a moral duty. *J Med Ethics 31*: 242–248.

Harris J, Holm S (2002) Extending human lifespan and the precautionary paradox. *J Medicine Philos 27*: 355–368.

Harris ML, Fufa TD, Palmer JW, Joshi SS, Larson DM, Incao A, Gildea DE, Trivedi NS, Lee AN, Day CP, Michael HT, Hornyak TJ, Merlino G, NISC Comparative Sequencing Program, Pavan WJ (2018) A direct link between MITF, innate immunity, and hair graying. *PLoS Biol* https://doi.org/10.1371/journal.pbio.2003648.

Harvey A, Brand A, Holgate ST, Kristiansen LV, Lehrach H, Palotie A, Prainsack B (2012) The future of technologies for personalised medicine. *New Biotechnology 29*: 625–633.

Hawkes K (2003) Grandmothers and the evolution of human longevity. *Am J Hum Biol 15*: 380–400.

Hawks J, Hunley K, Lee SH, Wolpoff M (2000) Population bottlenecks and Pleistocene human evolution. *Molecular Biology and Evolution 17*: 2–22.

Hayflick L, Moorhead PS (1961) The serial cultivation of human diploid cell strains. *Exp Cell Res* 1961; *25*: 562–585.

Hazan R, Que YA, Maura D, Strobel B, Majcherczyk PA, Hopper LR, Wilbur DJ, Hreha TN, Barquera B, Rahme LG (2016) Auto poisoning of the respiratory chain by a quorum-sensing-regulated molecule favors biofilm formation and antibiotic tolerance. *Curr Biol 26*: R80–R82.

Hedrick SM, Ch'en IL, Alves BN (2010) Intertwined pathways of programmed cell death in immunity. *Immunological Reviews 236*: 41–53.

Hegde NR, Maddur MS, Kaveriand SV, Bayry J (2009) Reasons to include viruses in the tree of life. *Nat Rev Microbiol 7*: 615.

Heron M (2019) Deaths: Leading causes for 2017. *National Vital Statistics* Reports Centers for Disease and Prevention. *68*(6).

Hernandez-Martinez R, Covarrubias L (2011) Interdigital cell death function and regulation: new insights on an old programmed cell death model. *Dev Growth Differ 53*: 245–258.

Hershkovitz I, Donoghue HD, Minnikin DE, Besra GS, Lee OY-C, Gernaey AM, Galili E, Eshed V, Greenblatt CL, Lemma E, Bar-Gal GK, Spigelman M (2008) Detection and molecular characterization of 9000-year-old *Mycobacterium tuberculosis* from a Neolithic settlement in the Eastern Mediterranean. *PLoS One 3*(10): e3426.

Hershkovitz I, Weber GW, Quam R, Duval M, Grün R, Kinsley L, Ayalon A, Bar-Matthews M, Valladas H, Mercier N, Arsuaga JL, Martinón-Torres J, Bermúdez de Castro JM, Fornai C, Martín-Francés L, Sarig R, May H, Krenn VA, Slon V, Rodríguez L, Garcia R, Lorenzo C, Carretero JM, Frumkin A, Shahack-Gross R, Bar-Yosef Mayer DE, Cui Y,

Wu X, Peled N, Groman-Yaroslavski I, Weissbrod L, Yeshurun R, Tsatskin A, Zaidner Y, Weinstein-Evron M (2018) The earliest modern humans outside Africa. *Science 359*: 456–459.

Hetzler PT 3rd, Dugdale LS (2018) How do medicalization and rescue fantasy prevent healthy dying? *AMA J Ethics 20*(8): E766–773.

Hipp MS, Kasturi P, Hartl FU (2019) The proteostasis network and its decline in ageing. *Nat Rev Mol Cell Biol 20*: 421–435.

Hitt R, Young-Xu Y, Silver M, Perls T (1999) Centenarians: the older you get, the healthier you have been. *The Lancet 354*: 652.

Hofreiter M, Stewart J (2009) Ecological change, range fluctuations and population dynamics during the Pleistocene. *Curr Biol 19*: R584–R594.

Holland S (2017) Treatment decision, death and the value of life. *QJM: An International Journal of Medicine 110*: 121–123.

Holmes TH, Rahe RH (1967) The social readjustment rating scale. *Journal of Psychosomatic Research 11*: 213–218.

Holstein TW, Laudet V (2014) Life-history evolution: At the origins of metamorphosis. *Curr Biol 24*: R159–R161.

Hoogendijk EO, Afilalo J, Ensrud KE, Kowal P, Onder G, Fried LP (2019) Frailty: implications for clinical practice and public health. *The Lancet 394*: 1365–1375.

Hoshika S, Leal NA, Kim MJ, Kim MS, Karalkar NB, Kim HJ, Bates AM, Watkins Jr NE, SantaLucia HA, Meyer AJ, DasGupta S, Piccirilli JA, Ellington AD, SantaLucia Jr J, Georgiadis MM, Benner SA (2019) Hachimoji DNA and RNA: A genetic system with eight building blocks. *Science 363*: 884–887.

Hotez PJ (2016) Neglected tropical diseases in the Anthropocene: The cases of Zika, Ebola, and other infections. *PLoS Negl Trop Dis 10*(4): e0004648.

Hotez PJ (2018) Southern Europe's coming plagues: Vector-borne neglected tropical diseases. *PLoS Negl Trop Dis 10*(6): e0004243.

Howell L, editor (2013) *World Economic Forum. Global risks 2013, eighth edition: an initiative of the Risk Response Network*. http://www3.weforum.org/docs/WEF_GlobalRisks_Report_2013.pdf. Retrieved on January 18, 2021.

Howells WW (1976) Explaining modern man: Evolutionists versus migrationists. *J Hum Evol 5*: 477–496.

Hoyle F, Wickramasinghe NC (1977) Polysaccharides and infrared spectra of galactic sources. *Nature 268*: 610–612.

Hublin J, Ben-Ncer A, Bailey S, Freidline SE, Neubauer S, Skinner MM, Bergmann I, Le Cabec A, Benazzi S, Harvati K, Gunz P (2017) New fossils from Jebel Irhoud, Morocco and the pan-African origin of *Homo sapiens*. *Nature 546*: 289–292.

Huff CD, Xing J, Rogers AR, Witherspoon D, Jorde LB (2010) Mobile elements reveal small population size in the ancient ancestors of *Homo sapiens*. *Proc Natl Acad Sci USA 107*: 2147–2152.

Hui D, dos Santos R, Chisholm G, Bansal S, Silva TB, Kilgore K, Crovador CS, Yu X, Swartz MD, Perez-Cruz PE, de Almeida Leite R, De Angelis Nascimento MS, Reddy S, Seriaco F, Yennu S, Paiva CE, Dev R, Hall S, Fajardo J, Bruera E (2014) Clinical signs of impending death in cancer patients. *The Oncologist 19*: 681–687.

Hull RS, Bjork VCE, Roy AG (2015) It is time to classify biological aging as a disease. *Frontiers in Genetics 6*: 205. doi: 10.3389/fgene.2015.00205.

Hunter MC, Pozhitkov AE, Noble PA (2017) Accurate predictions of postmortem interval using linear regression analyses of gene meter expression data. *Forensic Science International 275*: 90–101.

Hyde ER, Haarmann DP, Petrosino JF, Lynne AM, Bucheli SR (2015) Initial insights into bacterial succession during human decomposition. *Int J Legal Med 129*: 661–671.

Hyun I (2010) The bioethics of stem cell research and therapy. *J Clin Invest 120*: 71–75.

Ishizuya-Oka A, Hasebe T, Shi YB (2010) Apoptosis in amphibian organs during metamorphosis. *Apoptosis 15*: 350–364.

Jabr F (2017) How does the flu actually kill people? *Scientific American.* https://www. scientificamerican.com/article/how-does-the-flu-actually-kill-people/.

Jacobs GS, Hudjashov G, Saag L, Kusuma P, Carusallam CC, Lawson DJ, Mondal M, Pagani L, Ricaut FX, Stoneking M, Metspalu M, Sudoyo H, Lansing JS, Cox MP (2019) Multiple deeply divergent Denisovan ancestries in Papuans. *Cell 177*: 1–12.

JAMA (1899) Prevention of premature burial. *JAMA 32*(17): 947.

Jasilionis D, Shkolnikov VM, Andreev EM, Jdanov DA, Vågerö D, Meslé F, Vallin J (2012) Do vanguard populations pave the way towards higher life expectancy for other population groups? *Population 69*: 531–556.

Jiménez-Brobeil SA, Oumaoui A (2009) Possible relationship of cranial traumatic injuries with violence in the South-East Iberian Peninsula from the Neolithic to the Bronze Age. *Am J Phys Anthropol 140*: 465–475.

Jinek M, Chylinski K, Fonfara I, Hauer M, Doudna JA, Charpentier E (2012). A programmable dual-RNA-guided DNA endonuclease in adaptive bacterial immunity. *Science 337*: 816–821.

Joachimski MM, Lai X, Shen S, Jiang H, Luo G, Chen B, Chen J, Sun Y (2012) Climate warming in the latest Permian and the Permian–Triassic mass extinction. *Geology 40*, 195–198.

Johanson DC, Lovejoy CO, Kimbel WH, White TD, Ward SC, Bush ME, Latimer BM, Coppens Y (1982) Morphology of the Pliocene partial hominid skeleton (A.L. 288-1) from the Hadar Formation, Ethiopia. *American Journal of Physical Anthropology 57*: 403–451.

Joly Y, Dyke SOM, Knoppers BM, Pastinen T (2016) Are data sharing and privacy protection mutually exclusive? *Cell 167*: 1150–1154.

Jones DG, Dang JL (2006) The plant immune system. *Nature 444*: 323–329.

Jones DS, Podolsky SH, Greene JA (2012) The burden of disease and the changing task of medicine. *N Engl J Med 366*: 2333–2338.

Jones OR, Scheuerlein A, Salguero-Gómez, Camarda CG, Schaible R, Casper BB, Dahlgren JP, Ehrlén J, García MB, Menges ES, Quintana-Ascencio PF, Caswell H, Baudisch A, Vaupel JW (2014) Diversity of ageing across the tree of life. *Nature 505*: 169–173.

Kaczanowski S, Sajid M, Reece SE (2011) Evolution of apoptosis-like programmed cell death in unicellular protozoan parasites. *Parasites & Vectors 4*: 44.

Kaneda T, Haub C (2020) *How many people have ever lived on Earth?* Population Research Bureau. Retrieved from: https://www.prb.org/howmanypeoplehaveeverlivedonearth/.

Kanzaria HK, Probst MA, Hsia RY (2016) Emergency department death rates dropped by nearly 50 percent, 1997–2011. *Health Affairs (Millwood) 35*: 1303–1308.

Kappelman J, Alçiçek MC, Kazanci N, Schultz M, Ozkul M, et al. (2008) First *Homo erectus* from Turkey and implications for migrations into temperate Eurasia. *Am J Phys Anthropol 135*: 110–116.

Kappelman J, Ketcham RA, Pearce S, Todd L, Akins W, Colbert MW, Feseha M, Maisano JA, Witzel A (2016) Perimortem fractures in Lucy suggest mortality from fall out of tall tree. *Nature 537*: 503–507.

Karmin M. et al. (2015) A recent bottleneck of Y chromosome diversity coincides with a global change in culture. *Genome Res 25*: 459–466.

Katz BP, Zdeb MS, Therriault GD (1979) Where people die. *Public Health Report 94*: 522–527.

Kaufman S (2004) Autonomous agent. In: *Science and Ultimate Reality* (eds. JD Barrow, PCW Davis, CL Harper, Jr) Cambridge University Press. Cambridge.

Kaufman SR, Shim JK, Russ AJ (2006) Old age, life extension, and the character of medical choice. *J Gerontol 61B*: S175–S184.

Kehler DS (2019) Addressing social inequalities for longevity and living in good health. *Lancet Public Health 5*: e1–e70.

Kennedy BK, Berger SL, Brunet A, Campisi J, Cuervo AM, Epel ES, Franceschi C, Lithgow GJ, Morimoto RI, Pessin JE, Rando TA, Richardson A, Schadt EE, Wyss-Coray T, Sierra F (2014) Linking aging to chronic disease. *Cell 159*: 709–713.

Kennedy C, Bastiaens MT, Willemze R, Bavinck JNB, Bajdik CD, Westendorp RGJ (2003) Effect of smoking and sun on the aging skin. *Journal of Investigative Dermatology 120*: 548–554.

Kenyon C (2011) The first long-lived mutants: discovery of the insulin/IGF-1 pathway for ageing. *Philos Trans R Soc Lond B Biol Sci 366*: 9–16.

Kenyon C (2010) The genetics of ageing. *Nature 464*: 504–512.

Kerr JF, Wyllie AH, Currie AR (1972) Apoptosis: a basic biological phenomenon with wide-ranging implications in tissue kinetics. *Br J Cancer 126*: 239–257.

Khurana SM, Pandey SK, Sarkar D, Chanemougasoundharam A (2005) Apoptosis in plant disease response: A close encounter of the pathogen kind. *Current Science 88*: 740–752.

Kilpatrick AM, Randolph SE (2012) Drivers, dynamics, and control of emerging vector-borne zoonotic diseases. *The Lancet 380*: P1946–1955.

King G, Soneji S (2011) The future of death in America. *Demogr Res 25*: 1–38.

Kirkwood TBL (2005) Understanding the odd science of aging. *Cell 120*: 437–477.

Kirkwood TBL (2008) Understanding ageing from an evolutionary perspective. *J Intern Med 263*: 117–127.

Kirkwood TBL, Austad SN (2000) Why do we age? *Nature 408*: 233–238.

Kirkwood TBL, Holliday R (1979) The evolution of ageing and longevity. *Proc R Soc Lond B Biol Sci 205*: 531–546.

Klaus HD (2018) Possible prostate cancer in northern Peru: Differential diagnosis, vascular anatomy, and molecular signaling in the paleopathology of metastatic bone disease. *Intl J Paleopathol 21*: 147–157.

Klemenc-Ketis Z, Kersnik J, Grmec S (2010) The effect of carbon dioxide on near-death experiences in out-of-hospital cardiac arrest survivors: a prospective observational study. *Critical Care 14*: R56.

Knight J, Schilling C, Barnett A, Jackson R, Clarke P (2016) Revisiting the "Christmas Holiday Effect" in the Southern Hemisphere. *J Am Heart Assoc 5*:e005098.

Knott GJ, Doudna JA (2018) CRISPR-Cas guides the future of genetic engineering. *Science 361*: 866–869.

Kochanek KD, Murphy SL, Xu J, Arias E (2019) *Deaths: Final Data for 2017*. National Vital Statistics Reports 68, 9 https://www.cdc.gov/nchs/data/nvsr/nvsr68/nvsr68_09-508.pdf.

Kolodkin-Gal I, Hazan R, Gaathon A, Carmeli S, Engelberg-Kulka H (2007) A linear pentapeptide is a quorum-sensing factor required for mazEF-mediated cell death in *Escherichia coli*. *Science 318*: 652–655.

Komatsu K, Hashimoto M, Ozeki J, Yamaji Y, Maejima K, Senshu H, Himeno M, Okano Y, Kagiwada S, Namba S (2010) Viral-induced systemic necrosis in plants involves both programmed cell death and the inhibition of viral multiplication, which are regulated by independent pathways. *Molecular Plant-Microbe Interactions 23*: 283–293.

Kontis V, Bennett JE, Mathers CD, Li G, Foreman K, Ezzati M (2017) Future life expectancy in 35 industrialised countries: projections with a Bayesian model ensemble. *The Lancet 389*: 1323–1335.

Koshland DE (2002) The seven pillars of life. *Science 295*: 2215–2216.

Krammer PH, Arnold R, Lavrik IN (2007) Life and death in peripheral T cells. *Nat Rev Immunol 7*: 532–542.

Kreamer D (2013) The past, present, and future of water conflict and international security. *Journal of Contemporary Water Research & Education 149*: 88–96.

Krichbaum JG (1882) *Device for indicating life in buried persons.* US Patent 268,693. https://catalog.archives.gov/id/6277693 and https://patents.google.com/patent/US268693A/en. Both retrieved on January 18, 2021.

Kriegman S, Blackiston D, Levin M, Bongard J (2020) A scalable pipeline for designing reconfigurable organisms. *Proc Natl Acad Sci USA 117*: 1853–1859.

Kriete A (2013) Robustness and aging-a systems-level perspective. *Biosystems 112*: 37–48.

Kroemer G, Galluzzi L, Vandenabeele P, Adams J, Alnemri ES, Baehrecke EH, Blagosklonny MV, El-Deiry WS, Golstein P, Green DR, Hengartner M, Knight RA, Kumar S, Lipton SA, Malorni W, Nuñez G, Peter ME, Tschopp J, Yuan J, Piacentini M, Zhivotovsky B, Melino G (2009) Classification of cell death. *Cell Death Differ 16*: 3–11.

Krupovic M, Cvirkaite-Krupovic V (2011) Virophages or satellite viruses? *Nat Rev Microbiol 9*: 762–763.

Kuhn SL, Stiner MC (2006) What's a mother to do? The division of labor among Neandertals and modern humans in Eurasia. *Current Anthropology 47*: 953–981.

Laffel L, Svoren B (n.d.) *Epidemiology, presentation, and diagnosis of type 2 diabetes mellitus in children and adolescents.* Up to Date. https://www.uptodate.com/contents/epidemiology-presentation-and-diagnosis-of-type-2-diabetes-mellitus-in-children-and-adolescents.

Lai Z, Wang F, Zheng Z, Fan B, Chen Z (2011) A critical role of autophagy in plant resistance to necrotrophic fungal pathogens. *The Plant Journal 66*: 953–968.

Lambrechts L, Scott TW, Gubler DJ (2010) Global distribution of *Aedes albopictus* for dengue virus transmission. *PLoS Negl Trop Dis 4*(5): e646.

Landers J (1987) Mortality and metropolis: The case of London, 1675–1825. *Population Studies 41*: 59–76.

Landers J, Mouzas A (1988) Burial seasonality and causes of death in London, 1670–1819. *Population Studies 42*: 59–83.

Lane B, Gullone E (1999) Common fears: A comparison of adolescents' self-generated and fear survey schedule generated fears. *Journal of Genetic Psychology 160*: 194–204.

Lansky SB, List MA, Lansky LL, Ritter-Sterr C, Miller DR (1987) The measurement of performance in childhood cancer patients. *Cancer 60*: 1651–1656.

La Scola B, Audic S, Robert C, Jungang L, de Lamballerie X, Drancourt M, Birtles R, Claverie JM, Raoult D (2003) A giant virus in amoebae. *Science 299*: 2033.

Lau D, Ogbogu U, Taylor B, Stafinski T, Menon D, Caulfield T (2008) Stem cell clinics on-line: the direct-to-consumer portrayal of stem cell medicine. *Cell Stem Cell 3*: 591–594 (2008).

Laxminarayan R, Duse A, Wattal C, Zaidi AKM, Wertheim HFL, Sumpradit N, Vlieghe E, Levy Hara G, Gould IM, Goossens H, Greko C, So AD, Bigdeli M, Tomson G, Woodhouse W, Ombaka E, Peralta AQ, Qamar FN, Mir F, Kariuki S, Bhutta ZA, Coates A, Bergstrom R, Wright GD, Brown ED, Cars O (2013) Antibiotic resistance—the need for global solutions. *Lancet Infect Dis 13*: 1057–1098.

Leach DR, Krummel MF, Allison JP (1996) Enhancement of antitumor immunity by CTLA-4 blockade. *Science 271*: 1734–1736.

Lee SC, Pervaiz S (2007) Apoptosis in the pathophysiology of diabetes mellitus. *International Journal of Biochemistry & Cell Biology 39*: 497–504.

Lemke PA (1976) Viruses of eukaryotic microorganisms. *Annu Rev Microbiol 30*: 105–145.

Lenggenhager B, Tadi T, Metzinger T, Blanke O (2007) Video ergo sum: Manipulating bodily self-consciousness. *Science 317*: 1096–1099.

Leslie M (2001) Aging research grows up. *Sci. Aging Knowledge Environ 2002*(1): oa1.

Levy SB, Marshall B (2004) Antibacterial resistance worldwide: causes, challenges and responses. *Nat Med 10*: S122–S129.

Li J, Cao F, Yin HL, Huang ZJ, Lin ZT, Mao N, Sun B, Wang G (2020) Ferroptosis: Past, present and future. *Cell Death Dis.* 11: 88.

Li M, Wang H, Liu J, Hao P, Ma L, Liu Q (2016) The apoptotic role of metacaspase in *Toxoplasma gondii*. *Front Microbiol.* https://doi.org/10.3389/fmicb.2015.01560. Retrieved on January 18, 2021.

Libina N, Berman JR, Kenyon C (2003) Tissue-specific activities of *C. elegans* DAF-16 in the regulation of lifespan. *Cell 115*: 489–502.

Lindahl T, Barnes DE (2000) Repair of endogenous DNA damage. *Cold Spring Harb Symp Quant Biol 65*: 127–133.

Liu X, Huang J, Chen T, Wang Y, Xin S, Li J, Pei G, Kang J (2008) Yamanaka factors critically regulate the developmental signaling network in mouse embryonic stem cells. *Cell Research 18*: 1177–1189.

Ljubuncic P, Reznick AZ (2009) The evolutionary theories of aging revisited—A mini-review. *Gerontology 55*: 205–216.

Lockshin RA (2016) Programmed cell death 50 (and beyond). *Cell Death Differ 23*: 10–17.

Loike JD, Pollack R (2019) Opinion: How to define life. *The Scientist. May 2.* https://www.the-scientist.com/news-opinion/opinion--how-to-define-life-65831.

Longo VD, Panda S (2016) Fasting, circadian rhythms, and time restricted feeding in healthy lifespan. *Cell Metab 23*: 1048–1059.

Longrich NR (2012) Mass extinction of lizards and snakes at the Cretaceous–Paleogene boundary. *Proc Natl Acad Sci USA 109*: 21396–21401.

López-León M, Goya RG (2017) The emerging view of aging as a reversible epigenetic process. *Gerontology 63*: 426–431.

López-Otín C, Blasco MA, Partridge L, Serrano M, Kroemers G (2013) The hallmarks of aging. *Cell 153*: 1194–1217.

Lu BC, Gallo N, Kues U (2003) White-cap mutants and meiotic apoptosis in the basidiomycete *Coprinus cinereus*. *Fungal Genetics and Biology 39*: 82–93.

Lu BCK (2006) Programmed cell death in fungi. In: *The Mycota I Growth, Differentiation and Sexuality* (Kües U, Fischer R, eds), pp. 167–187. Springer, Berlin.

Lucey BP, Nelson-Rees WA, Hutchins GM (2009) Henrietta Lacks, HeLa cells, and cell culture contamination. *Arch Pathol Lab Med 133*: 1463–1467.

Lucke JC, Diedrichs PC, Partridge B, Hall WD (2009) Anticipating the anti-ageing pill. *EMBO Rep 10*: 108–113.

Lucke JC, Hall W (2005) Who wants to live forever? *EMBO Rep* 6: 98–102.

Lund J, Tedesco P, Duke K, Wang J, Kim SK, Johnson TE (2002) Transcriptional profile of aging in C. elegans. *Curr Biol* 12: 1566–1573.

Lyman RL (2010) What taphonomy is, what it isn't, and why taphonomists should care about the difference. *Journal of Taphonomy* 8: 1–16.

MacDonald SWS, Hultsch DF, Dixon RA (2011) Aging and the shape of cognitive change before death: Terminal decline or terminal drop? *The Journals of Gerontology: Series B* 66B: 292–301.

MacDougall D (1907) The Soul: Hypothesis concerning soul substance together with experimental evidence of the existence of such substance. *American Medicine* 2: 240–243.

Maheshwari R, Navaraj A (2007) Senescence in fungi: the view from *Neurospora*. *FEBS Microbiol Lett* 280: 135–143.

Mahley RM (2017) Apolipoprotein E: Remarkable protein sheds light on cardiovascular and neurological diseases. *Clinical Chemistry* 63: 14–20.

Maklalov AA, Rowe L, Friberg U (2015) Why organisms age: Evolution of senescence under positive pleiotropy? *Bioessays* 37: 802–807.

Marcon AR, Murdoch B, Caulfield T (2017) Fake news portrayals of stem cells and stem cell research. *Regen Med* 12: 765–775.

Marean CW, Bar-Matthews M, Fisher E, Goldberg P, Herries A, Karkanas P, Nilssen PJ, Thompson E (2010) The stratigraphy of the Middle Stone Age sediments at Pinnacle Point Cave 13B (Mossel Bay, Western Cape Province, South Africa). *J Hum Evol* 59: 234–255.

Margulis L, Sagan D (2000) *What Is Life?* University of California Press, Berkeley, CA.

Martial C, Cassol H, Antonopoulos G, Charlier T, Heros J, Donneau AF, Charland-Verville V, Laureys S (2017) Temporality of features in near-death experience narratives. *Front Hum Neurosci* 11: 311.

Martial C, Cassol H, Charland-Verville V, Merckelbach H, Laureys S (2018) Fantasy proneness correlates with the intensity of near-death experience. *Front Psychiatry* 9: 190.

Martignoni ME, Kunze P, Friess H (2003) Cancer cachexia. *Mol Cancer* 2: 36.

Martin GM (2007) Modalities of gene action predicted by the classical evolutionary biological theory of aging. *Ann NY Acad Sci* 1100: 14–20.

Martin RL, Lloyd HE, Cowan AI (1994) The early events of oxygen and glucose deprivation: setting the scene for neuronal death? *Trends in Neurosciences* 17: 251–257.

Matthews T, Wilby RL, Murphy C (2019) An emerging tropical cyclone-deadly heat compound hazard. *Nat Clim Change* 9: 602–606.

Maturana H, Varela F (1980) *Autopoiesis and Cognition: The Realization of the Living* (2nd edition). Springer Netherlands.

Maynard S, Fang EF, Scheibye-Knudsen M, Croteau DL, Bohr VA (2015) DNA damage, DNA repair, aging, and neurodegeneration. *Cold Spring Harb Perspect Med* 5: a025130.

Mazarakis ND, Edwards AD, Mehmet H (1997) Apoptosis in neural development and disease. *Archives of Disease in Childhood* 77: F165–F170.

McAuley MT, Martinez Guimera A, Hodgson D, Mcdonald N, Mooney KM, Morgan AE, Proctor CJ (2017) Modelling the molecular mechanisms of aging. *Bioscience Reports* 37: BSR20160177.

McCarthy JV (2003) Apoptosis and development. *Essays in Biochemistry* 39: 11–24.

McClearn GE, Johansson B, Berg S, Pedersen NL, Ahern F, Petrill SA, Plomin R (1997) Substantial genetic influence on cognitive abilities in twins 80 or more years old. *Science* 276: 1560–1563.

McConnel C, Turner L (2005) Medicine, ageing and human longevity. *EMBO Rep* 6: S59–S62.

McDonald MM, Navarrete CD, Van Vugt M (2012) Evolution and the psychology of intergroup conflict: the male warrior hypothesis. *Phil. Trans. R. Soc. B 367*: 670–679.

McMenamin SK, Parichy DM (2013) Metamorphosis in teleosts. *Curr Top Dev Biol 103*: 127–165.

McMichael AJ, Lindgren E (2011) Climate change: present and future risks to health, and necessary responses. *J Intern Med 270*: 401–413.

Medawar PB (1952) *An Unsolved Problem of Biology*. London, HK Lewis.

Medvedev ZA (1990) An attempt at a rational classification of theories of ageing. *Biol Rev Camb Philos Soc 65*: 375–398.

Medzhitov R, Janeway CA Jr (2002) Decoding the patterns of self and nonself by the innate immune system. *Science 296*: 298–300.

Merideth MA, Gordon LB, Clauss S, Sachdev V, Smith ACM, Perry MB, Brewer CC, Zalewski C, Kim HJ, Solomon B, Brooks BP, Gerber LH (2008) Phenotype and course of Hutchinson–Gilford progeria syndrome. *N Engl J Med* 2008; *358*: 592–604.

Merksamer PI, Liu Y, He W, Hirschey MD, Chen D, Verdin E (2013) The sirtuins, oxidative stress and aging: an emerging link. *Aging (Albany NY) 5*: 144–150.

Michaeli S, Galili G, Genschik P, Fernie AR, Avin-Wittenbert TA (2015) Autophagy in plants—What's new on the menu? *Trends in Plant Science 21*: 134–144.

Michaud M, Balardy L, Moulis G, Gaudin C, Peyrot C, Vellas B, Cesari M, Nourhashemi F (2013) Proinflammatory cytokines, aging, and age-related diseases. *J Am Med Dir Assoc 14*: 877-882.

Miller G (2005) What is the biological basis of consciousness? *Science 309*: 79.

Miller SL, Urey HC (1959) Organic compound synthesis on the primitive Earth. *Science 130*: 245–251.

Minina EA, Bozhkov PV, Hofius D (2014) Autophagy as initiator or executioner of cell death. *Trends in Plant Science 19*: 692–697.

Mitani JC, Watts DP, Amsler SJ (2010) Lethal intergroup aggression leads to territorial expansion in wild chimpanzees. *Curr Biol 20*: R507–R508.

Mitchell LA, McCulloch LH, Pinglaya S, Bergera H, Bosco N, Brosha R, Bulajicd M, Huanga E, Hogan MS, Martin JA, Mazzonid EO, Davolia T, Maurano MT, Boekea JD (2019) De novo assembly, delivery and expression of a 101 kb human gene in mouse cells. *BioRxiv* doi: https://doi.org/10.1101/423426.

Mitchell RN, Kilian TM, Evans DAD (2012) Supercontinent cycles and the calculation of absolute palaeolongitude in deep time. *Nature 482*: 208–212.

Mitler MM, Hajdukovic RM, Shafor R, Hahn PM, Kripke DF (1987) When people die. *Am J Med 82*: 266–274.

Mitteldorf J, Fahy GM (2018) Questioning the inevitability of aging. *Proc Natl Acad Sci USA 115*: E558.

Mobbs D, Watt C (2011) There is nothing paranormal about near-death experiences: how neuroscience can explain seeing bright lights, meeting the dead, or being convinced you are one of them. *Trends in Cognitive Sciences 15*: 447–449.

Mollaret P, Goulon M (1959) The depassed coma. *Rev Neurol 101*: 3–15.

Monge J, Kricum M, Radovcic J, Radovcic D, Mann A, Frayer DW (2013) Fibrous dysplasia in a 120,000+ year old Neandertal from Krapina, Crotia. *PLoS One 8*(6): e64539.

Moore DJ, West AB, Dawson VL, Dawson TM (2005) Molecular pathophysiology of Parkinson's disease. *Annu Rev Neurosci 28*: 57–87.

Mora C, Dousset B, Caldwell IR, Powell FE, Geronimo RC, Bielecki CR, Counsell CWW, Dietrich BS, Johnston ET, Louis LV, Lucas MP, McKenzie MM, Shea AG, Tseng H, Giambelluca TW, Leon LR, Hawkins E, Trauernicht C (2017) Global risk of deadly heat. *Nat Clim Change* 7: 501–506.

Moreira D, Lopez-Garcia P (2009) Ten reasons to exclude viruses from the tree of life. *Nat Rev Microbiol* 7: 306–331.

Morens DM, Fauci AS (2013) Emerging infectious diseases: Threats to human health and global stability. *PLoS Pathog* 9(7): e1003467.

Morrill S, He DZZ (2017) Apoptosis in inner ear sensory hair cells. *Journal of Otology* 12: 151–164.

Mortensen M, Soilleux EJ, Djordjevic G, Tripp R, Lutteropp M, Sadighi-Akha E, Stranks AJ, Glanville J, Knight S, Jacobsen S-EW, Kranc KR, Simon AK (2011) The autophagy protein Atg7 is essential for hematopoietic stem cell maintenance. *J Exp Med* 208: 455–467.

Moskalev AA, Shaposhnikov MV, Plyusnina EN, Zhavoronkov A, Budovsky A, Yanai H, Fraifeld VE (2012) The role of DNA damage and repair in aging through the prism of Koch-like criteria. *Ageing Res Rev* 12: 661–684.

Mowat F (2010) *Never Cry Wolf.* Logan, IA: Perfection Learning.

Mulvihill JJ, Capps B, Joly Y, Lysaght T, Zwart HAE, Chadwick R, International Human Genome Organization Committee of Ethics, Law and Society (2017) Ethical issues of CRISPR technology and gene editing through the lens of solidarity. *Br Med Bull* 122: 17–29.

Muraguchi H, Fujita T, Kishibe Y, Konno K, Ueda N, Nakahori K, Yanagi SO, Kamada T (2008) The *exp1* gene essential for pileus expansion and autolysis of the inky cap mushroom *Coprinopsis cinerea* (*Coprinus cinereus*) encodes an HMG protein. *Fungal Genetics and Biology* 45: 890–896.

Murphy KM, Topel RH (2003) Diminishing returns? The costs and benefits of improving health. *Perspect Biol Med* 46: S108–S128.

NA (nd) *What you need to know about disease. Disease threats. Global killers.* The National Academies. http://needtoknow.nas.edu/id/threats/global-killers/.

Nakajima K, Fujimoto K, Yaoita Y (2005) Programmed cell death during amphibian metamorphosis. *Seminars in Cell & Developmental Biology* 16: 271–280.

Napier WM (2015) Giant comets and mass extinctions of life. *Monthly Notices of the Royal Astronomical Society* 448: 27–36.

NASA (2020) *About life detection. Astrobiology at NASA. Life in the Universe.* https://astrobiology.nasa.gov/research/life-detection/about/. Retrieved on January 18, 2021.

NASA (2021) https://www.nasa.gov/vision/universe/starsgalaxies/life%27s_working_definition.html. Retrieved on November 30, 2021.

National Academies of Sciences Engineering and Medicine (US) (2017) *Human Genome Editing: Science, Ethics, and Governance,* The National Academies Press, Washington, DC.

Navas-Castillo J (2009) Six comments on the ten reasons for the demotion of viruses. *Nat Rev Microbiol* 7: 615–617.

NCI (2018) *Cancer statistics.* National Cancer Institute. https://www.cancer.gov/about-cancer/understanding/statistics. Retrieved on January 18, 2021.

Ndozangue-Touriguine O, Hamelin J, Breard J (2008) Cytoskeleton and apoptosis. *Biochem Pharmacol* 76: 11–18.

Neal JV (1962) Diabetes mellitus: A "thrifty" genotype rendered detrimental by "progress"? *Am J Hum Genet 14*: 353–362.

Nedjic J, Aichinger M, Emmerich J, Mizushima N, Klein L (2008) Autophagy in thymic epithelium shapes the T-cell repertoire and is essential for tolerance. *Nature 455*: 396–400.

Neiderud CJ (2015) How urbanization affects the epidemiology of emerging infectious diseases, *Infection Ecology & Epidemiology 5*: 1, 27060.

Nelson C, Baehrecke EH (2014) Eaten to death. *FEBS J 281*: 5411–5417.

Nelson P, Masel J (2017) Intercellular competition and the inevitability of multicellular aging. *Proc Natl Acad Sci USA 114*: 12982–12987.

Neu J, Walker WA (2011) Necrotizing enterocolitis. *N Engl J Med 364*: 255–264.

Neves J, Demaria M, Campisi J, Jasper H (2015) Of flies, mice, and men: Evolutionarily conserved tissue damage responses and aging. *Developmental Cell 32*: 9–18.

Newman SJ (2018) Errors as a primary cause of late-life mortality deceleration and plateaus. *PLoS Biol 16*(12): e2006776.

Ni XL, Su H, Zhou YF, Wang FH, Liu WZ (2015) Leaf-shape remodeling: programmed cell death in fistular leaves of *Allium fistulosum*. *Physiologia Plantarum 153*: 419–431.

Niccoli T, Partridge L (2012) Ageing as a risk factor for disease. *Curr Biol 22*: R741–R752.

Nielsen R, Akey JM, Jakobsson M, Pritchard JK, Tishkoff S, Willerslev ET (2017) Tracing the peopling of the world through genomics. *Nature 541*: 302–310.

Niño DF, Sodhi C, Hackam DJ (2016) Necrotizing enterocolitis: New insights into pathogenesis and mechanisms. *Nat Rev Gastroenterol Hepatol 13*: 590–600.

Nishimura H, Nose M, Hiai H, Minato N, Honjo T (1999) Development of lupus-like autoimmune diseases by disruption of the PD-1 gene encoding an ITIM motif-carrying immunoreceptor. *Immunity 11*: 141–151.

Noireaux V, Maeda YT, Libchaber A (2011) Development of an artificial cell, from self-organization to computation and self-reproduction. *Proc Natl Acad Sciences USA 108*: 3473–3480.

Norton RJ (nd) *Abraham Lincoln's body exhumed and viewed in 1901.* https://rogerjnorton.com/Lincoln13.html. Retrieved January 18, 2021.

NSC (2020) *What are the odds of dying from . . .* National Safety Council website. https://www.nsc.org/work-safety/tools-resources/injury-facts/chart.

Nunn N, Qian N (2010) The Columbian Exchange: A history of disease, food, and ideas. *Journal of Economic Perspectives 24*: 163–188.

Obukhova LA, Skulachev VP, Natalia G, Kolosova NG. Mitochondria- targeted antioxidant SkQ1 inhibits age-dependent involution of the thymus in normal and senescence-prone rats. *Aging*. 2009; 1: 389–401.

Ocampo A, Reddy P, Martinez-Redondo P, Platero-Luengo A, Hatanaka F, Hishida T, Li M, Lam D, Kurita M, Beyret E, Araoka T, Vasquez-Ferrer E, Donoso D, Roman JL, Xu J, Rodriguez Esteban C, Nuñez G, Nuñez Delicado E, Campistol JM, Guillen I, Guillen P, Belmonte JC (2016) In vivo amelioration of age-associated hallmarks by partial reprogramming. *Cell 167*: 1719–1733.

Odes EJ, Randolph-Quinney PS, Steyn M, Zach Throckmorton Z, Smilg JS, Zipfel B, Augustine TN, de Beer F, Hoffman JW, Franklin RD, Berge LR (2016) Earliest hominin cancer: 1.7-million-year old osteosarcoma from Swartkrans Cave, South Africa. *S Afr J Sci 112*(7/8): 100–104.

Oeppen J, Vaupel JW (2002) Broken limits to life expectancy. *Science 296*: 1029–1031.

Oken MM, Creech RH, Tormey DC, Horton J, Davis TE, McFadden ET, Carbone PP (1982) Toxicity and response criteria of the Eastern Cooperative Oncology Group. *Am J Clin Oncol 5*: 649–655.

Olshansky SJ, Passaro DJ, Hershow RC, Layden J, Carnes BA, Brody J, Hayflick L, Butler RN, Allison DB, Ludwig DS (2005) A potential decline in life expectancy in the United States in the 21st Century. *N Engl J Med 352*: 1138–1145.

Olson PR (2014) Flush and bone. Funeralizing alkaline hydrolysis in the United States. *Science, Technology, and Human Values 39*: 666–693.

Orvedahl A, Levine B (2009) Eating the enemy within: Autophagy in infectious diseases. *Cell Death Differ 16*: 57–69.

Owsley C (2011) Aging and vision. *Vision Res 51*: 1610–1622.

Pal S, Tyler JK (2016) Epigenetics and aging. *Science Advances 2*: e1600584.

Pani P, Giarrocco F, Giamundo M, Brunamonti E, Mattia M, Ferraina S (2018) Persistence of cortical neuronal activity in the dying brain. *Resuscitation 130*: e5–e7.

Pardoll DM (2012) The blockade of immune checkpoints in cancer immunotherapy. *Nat Rev Cancer 12*: 252–264.

Park SJ, Ahmad F, Philp A, Baar K, Williams T, Luo H, Ke H, Rehmann H, Taussig R, Brown AL, Kim MK, Beaven MA, Burgin AV, Manganiello V, Chung JH (2012) Resveratrol ameliorates aging-related metabolic phenotypes by inhibiting cAMP phosphodiesterases. *Cell 148*: 421–433.

Parks CL (2011) A study of the human decomposition sequence in Central Texas. *J Forensic Sci 56*: 19–22.

Partridge B, Hall W (2007) The search for Methuselah. Should we endeavor to increase the maximum human lifespan? *EMBO Rep 8*: 888–891.

Paul JY, Becker DK, Dickman MB, Harding RM, Khanna HK, Dale JL (2011) Apoptosis-related genes confer resistance to Fusarium wilt in transgenic 'Lady Finger' bananas. *Plant Biotechnology Journal 9*: 1141–1148.

Pawlikowska L, Hu D, Huntsman S, Sung A, Chu C, Chen J, Joyner AH, Schork NJ, Hsueh W-C, Reiner AP, Psaty BM, Atzmon G, Barzilai N, Cummings SR, Browner WS, Kwok P-Y, Ziv E (2009) Association of common genetic variation in the insulin/IGF1 signaling pathway with human longevity. *Aging Cell 8*: 460–472.

Payne JL, Clapham ME (2012) End-Permian mass extinction in the oceans: An ancient analog for the Twenty-First Century? *Annu Rev Earth Planet Sci 40*: 89–111.

Pearce-Duvet JMC (2006) The origin of human pathogens: evaluating the role of agriculture and domestic animals in the evolution of human disease. *Biological Reviews 81*: 369–382.

Penn Wharton (2016) *Mortality in the United States: Past, present, and future.* Wharton School, University of Pennsylvania, http://budgetmodel.wharton.upenn.edu/issues/2016/1/25/mortality-in-the-united-states-past-present-and-future. Retrieved on January 18, 2021.

Perri AR, Smith GM, Bosch MD, (2015) Comment on "How do you kill 86 mammoths? Taphonomic investigations of mammoth megasites" by Pat Shipman. *Quat Int 368*: 112e115.

Persat A, Nadell CD, Kim MK, Ingremeau F, Siryaporn A, Drescher K, Wingreen NS, Bassler BL, Gitai Z, Stone HA (2015) The mechanical world of bacteria. *Cell 161*: 988–997.

Péus D, Newcomb N, Hofer S (2013) Appraisal of the Karnofsky Performance Status and proposal of a simple algorithmic system for its evaluation. *BMC Med Inform Decis Mak.13*: 72.

Philipp O, Hamann A, Servos J, Werner A, Koch I, Osiewacz HD (2013) A genome-wide longitudinal transcriptome analysis of the aging model *Podospora anserine*. *PLoS ONE* 8: e83109.

Pijnenburg MAM, Leget C (2007) Who wants to live forever? Three arguments against extending the human lifespan. *J Med Ethics* 33: 585–587.

Piotrowska M (2019) Avoiding the potentiality trap: thinking about the moral status of synthetic embryos. *Monash Bioeth Rev* 38: 166–180.

Posth C, Wißing C, Kitagawa K, Pagani L, van Holstein L, Racimo F, Wehrberger K, Conard NJ, Kind CJ, Bocherens H, Krause J (2017) Deeply divergent archaic mitochondrial genome provides lower time boundary for African gene flow into Neanderthals. *Nat Commun* 8: 16046.

Powner MW, Gerland B, Sutherland JD (2009) Synthesis of activated pyrimidine ribonucleotides in prebiotically plausible conditions. *Nature* 459: 239–242.

Pozhitkov AE, Neme R, Domazet-Lošo T, Leroux BG, Soni S, Tautz D, Noble PA (2017) Tracing the dynamics of gene transcripts after organismal death. *Open Biol* 7: 160267.

Prates C, Sousa S, Oliveira C, Ikram S (2011) Prostate metastatic bone cancer in an Egyptian Ptolemaic mummy, a proposed radiological diagnosis. *Intl J Paleopathol* 1: 98–103.

President's Commission for the Study of Ethical Problems in Medicine and Biomedical and Behavioral Research (1981) *Defining death: a report on the medical, legal and ethical issues in the determination of death*. https://repository.library.georgetown.edu/handle/10822/559345. Retrieved on January 18, 2021.

President's Council on Bioethics (2003) *Beyond therapy: Biotechnology and the pursuit of happiness*. https://bioethicsarchive.georgetown.edu/pcbe/reports/beyondtherapy/. Retrieved on January 18, 2021.

Pretorius E, du Plooy JN, Bester J (2016) A comprehensive review on eryptosis. *Cell Physiol Biochem* 39: 1977–2000.

Proctor CJ, Lorimer IA (2011) Modelling the role of the Hsp70/Hsp90 system in the maintenance of protein homeostasis. *PLoS ONE* 6: e22038.

Proto WR, Coombs GH, Mottram JC (2013) Cell death in parasitic protozoa: regulated or incidental? *Nat Rev Microbiol* 11: 58–66.

Puca A, Daly MJ, Brewster SJ, Matise TC, Barrett J, Shea-Drinkwater M, Kang S, Joyce E, Nicoli J, Benson E, Kunkel LM, Perls T (2001) A genome-wide scan for linkage to human exceptional longevity identifies a locus on chromosome 4. *Proc Natl Acad Sci USA* 98: 10505–10508.

Pujolar JM, Jacobsen MW, Bekkevold D, Lobón-Cervià J, Jónsson B, Bernatchez L, Hansen MM (2015) Signatures of natural selection between life cycle stages separated by metamorphosis in European eel. *BMC Genomics* 16: 600.

Puleston DJ, Simon AK (2013) Autophagy in the immune system. *Immunology* 141: 1–8.

Putnam ML, Sindermann AB (1994) Eradication of potato wart disease from Maryland. *American Potato Journal* 71: 743–747.

Qian L, Huang Y, Spencer CI, Foley A, Vedantham V, Liu L, Conway SJ, Fu JD, Srivastava D (2012) In vivo reprogramming of murine cardiac fibroblasts into induced cardiomyocytes. *Nature* 485: 593–598.

Rabett RJ (2018) The success of failed *Homo sapiens* dispersals out of Africa and into Asia. *Nat Ecol Evol* 2: 212–219.

Ragsdale AP, Gravel S (2019) Models of archaic admixture and recent history from two-locus statistics. *PLoS Genet* 15(6): e1008204.

Ramisetty BCM, Natarajan B, Sarojini R (2013) *mazEF*-mediated programmed cell death in bacteria: "What is this?" *Crit Rev Microbiol* 41: 89–100.

Reape TJ, Molony EM, McCabe PF (2008) Programmed cell death in plants: Distinguishing between different modes. *J Exp Biol 59*: 435–444.

Redman JM, Gibney GT, Atkins MB (2016) Advances in immunotherapy for melanoma. *BMC Medicine 14*: 20.

Reich D, Green RE, Kircher M, Krause J, Patterson N, Durand EY, Viola B, Briggs AW, Stenzel U, Johnson PLF, Maricic T, Good JM, Marques-Bonet T, Alkan C, Fu Q, Mallick S, Li H, Meyer M, Eichler EE, Soneking M, Richards M, Talamo S, Shunko MV, Derevianko AP, Hublin JJ, Kelso J, Slatkin M, Paabo (2010) Genetic history of an archaic hominin group from Denisova Cave in Siberia. *Nature 468*: 1053–1060.

Remarque EM (1929) *All Quiet on the Western Front*. Little, Brown, New York.

Renne PR, Deino AL, Hilgen FJ, Kuiper KF, Mark DF, Mitchell III WS, Morgan LE, Mundil R, Smit J (2013) Time scales of critical events around the Cretaceous-Paleogene boundary. *Science 339*: 684–687.

Resnick SM, Pham DL, Kraut MA, Zonderman AB, Davatzikos C (2003) Longitudinal magnetic resonance imaging studies of older adults: a shrinking brain. *J Neurosci 23*: 3295–3301.

Reynolds RM, Temiyasathit S, Reedy MM, Ruedi EA, Drnevich JM, Leips J, Hughes KA (2007) Age specificity of inbreeding load in Drosophila melanogaster and implications for the evolution of late-life mortality plateaus. *Genetics 177*: 587–595.

Richter D, Grün R, Joannes-Boyau R, Steele TE, Amani F, Rué M, Fernandes P, Raynal JP, Geraads D, Ben-Ncer A, Hublin JJ, McPherron (2017) The age of the hominin fossils from Jebel Irhoud, Morocco, and the origins of the Middle Stone Age. *Nature 546*: 293–296.

Rigaud K, Kanta, de Sherbinin A, Jones B, Bergmann J, Clement V, Ober K, Schewe J, Adamo S, McCusker B, Heuser S, Midgley A (2018) *Groundswell: Preparing for internal climate migration*. The World Bank. https://openknowledge.worldbank.org/handle/10986/29461. Retrieved on January 18, 2021.

Rivron N, Pera M, Rossant J, Martinez Arias A, Zernicka-Goetz M, Fu J, van den Brink S, Bredenoord A, Dondorp W, de Wert G, Hyun I, Munsie M, Isasi R (2018) Debate ethics of embryo models from stem cells. *Nature 564*: 183–185.

Robine JM, Allard M, Herrmann FR, Jeune B (2019) The real facts supporting Jeanne Calment as the oldest ever human. *The Journals of Gerontology: Series A 74*: S13–20.

Robins C, Conneely KN (2014) Testing evolutionary models of senescence: Traditional approaches and future directions. *Hum Genet 133*: 1451–1465.

Robock A, Ammann CM, Oman L, Shindell D, Levis S, Stenchikov G (2009) Did the Toba volcanic eruption of ~74 ka B.P. produce widespread glaciation? *J Geophys Res 114*: D10107, doi: 10.1029/2008JD011652.

Rodier F, Campisi J (2011) Four faces of cellular senescence. *J Cell Biol 192*: 547–556.

Rolff J, Johnston PR, Reynolds S (2019) Complete metamorphosis of insects. *Philos Trans R Soc Lond B Biol Sci 374*: 20190063.

Rose MR (1984) Laboratory evolution of postponed senescence in Drosophila melanogaster. *Evolution 38*: 1004–1010.

Rowley PA, Ho B, Bushong S, Johnson A, Sawyer SL (2016) *XRN1* is a species-specific virus restriction factor in yeasts. *PLoS Pathog 12*(10): e1005890.

Rubin GM (1988) Drosophila melanogaster as an experimental organism. *Science 240*: 1453–1459.

Rubinsztein DC, Shpilka T, Elazar Z (2012) Mechanisms of autophagosome biogenesis. *Curr Biol 22*: R29–R34.

Rushby AJ, Claire MW, Osborn H, Watson AJ (2013) Habitable zone lifetimes of exoplanets around main sequence stars. *Astrobiology 13*: 833–849.

Sagan C (1997) *Billions and Billions: Thoughts on Life and Death at the Brink of the Millennium.* Random House, New York.

Sakamaki K, Nozaki M, Kominami K, Satou Y (2007) The evolutionary conservation of the core components necessary for the extrinsic apoptotic signaling pathway, in Medaka fish. *BMC Genomics 8*: 141.

Sakamoto Y, Nakade K, Konno N, Sato T (2012) Senescence of the *Lentinula edodes* fruiting body after harvesting. In: Kapiris, K (Ed.), *Food Quality.* InTech North America, New York, 83–110.

Sala N, Arsuaga JL, Pantoja-Pérez A, Pablos A, Martínez I, Quam RM, Asier Gómez-Olivencia A, José María Bermúdez de Castro JM, Carbonell E (2015) Lethal interpersonal violence in the Middle Pleistocene. *PLoS ONE 10*(5): e0126589.

Salles N (2007) Basic mechanisms of the aging gastrointestinal tract. *Dig Dis 25*: 112–117.

Sallon S, Solowey E, Cohen Y, Korchinsky R, Egli M, Woodhatch I, Simchoni O, Kislev M (2008) Germination, genetics, and growth of an ancient date seed. *Science 320*: 1464.

Sammut D, Craig C (2019) *Bodies in the Bog: The Lindow mysteries.* Science History Institute. https://www.sciencehistory.org/distillations/bodies-in-the-bog-the-lindow-mysteries. Retrieved on January 18, 2021.

Sankararaman S, Mallick S, Patterson N, Reich D (2016) The combined landscape of Denisovan and Neanderthal ancestry in present-day humans. *Curr Biol 26*: 1241–1247.

Sarbey B (2016) Definitions of death: brain death and what matters in a person. *Journal of Law and the Biosciences 3*: 743–752.

Saupe SJ (2000) Molecular genetics of heterokaryon incompatibility in filamentous ascomycetes. *Microbiol Mol Biol Rev 64*: 489–502.

Scaffidi P, Misteli T (2005) Reversal of the cellular phenotype in the premature aging disease Hutchinson-Gilford Progeria Syndrome. *Nat Med 11*: 440–445.

Schatz O, Langer E, Ben-Arie N (2014) Gene dosage of the transcription factor *Fingerin* (*bHLHA9*) affects digit development and links syndactyly to ectrodactyly. *Human Molecular Genetics 23*: 5394–5401.

Schiff ND (2010) Recovery of consciousness after brain injury: a mesocircuit hypothesis. *Trends in Neuroscience 33*: 1–9.

Schmid M, Simpson DJ, Sarioglu H, Lottspeich F, Gietl C (2001) The ricinosomes of senescing plant tissue bud from the endoplasmic reticulum. *Proceedings of the National Academy of Sciences USA 98*: 5353–5358.

Schmid-Siegert E, Sarkar N, Iseli C, Calderon S, Gouhier-Darimont C, Chrast J, Cattaneo P, Schütz F, Farinelli L, Pagni M, Schneider M, Voumard J, Jaboyedoff M, Fankhauser C, Hardtke CS, Keller L, Pannell JR, Reymond A, Robinson-Rechavi M, Xenarios I, Reymond P (2017) Low number of fixed somatic mutations in a long-lived oak tree. *Nat Plants 12*: 926–929.

Schneebeili-Hermann E (2012) Extinguishing a Permian World. *Geology 40*: 287–288.

Schoene B, Samperton KM, Eddy MP, Keller G, Adatte T, Bowring SA, Khadri SF, Gertsch B (2015) Earth history. U-Pb geochronology of the Deccan Traps and relation to the end-Cretaceous mass extinction. *Science 347*: 182–184.

Scholthof K-BG (2000) Tobacco mosaic virus. *The Plant Health Instructor.* DOI: 10.1094/PHI-I-2000-1010-01. Retrieved January 18, 2021.

Schramma AE, Carton-Leclercqa A, Dialloa S, Navarroa V, Chaveza M, Mahona S, Charpiera S (2020) Identifying neuronal correlates of dying and resuscitation in a model of reversible brain anoxia. *Progress in Neurobiology 185*: 101733.

Schrödinger E (1944) *What Is Life?* Cambridge University Press, Cambridge, UK.

Schulte P, et al. (2010) The Chicxulub asteroid impact and mass extinction at the Cretaceous-Paleogene boundary. *Science 327*: 1214–1218.

Schultz M, Parzinger H, Posdnjakov DV, Chikisheva TA, Schmidt-Schultz TH (2007) Oldest known case of metastasizing prostate carcinoma diagnosed in the skeleton of a 2,700-year-old Scythian King from Arzhan (Siberia, Russia). *Int J Cancer 121*: 2591–2595.

Schulz F, Yutin N, Ivanova NN, Ortega DR, Lee TK, Vierheilig J, Daims H, Horn M, Wagner M, Jensen GJ, Kyrpides NC, Koonin EV, Woyke T (2017) Giant viruses with an expanded complement of translation system components. *Science 356*: 82–85.

Sen P, Shah PP, Nativio R, Berger SL (2016) Epigenetic mechanisms of longevity and aging. *Cell 166*: 822–839.

Shah SK, Truog RD, Miller FG (2011) Death and legal fictions. *J Med Ethics 37*: 705.

Sharon A, Finkelstein A, Shlezinger N, Hatam I (2009) Fungal apoptosis: function, genes and gene function. *FEMS Microbiol Rev 33*: 833–854.

Sheehan PM (2001) The late Ordovician mass extinction. *Annu Rev Earth Planet Sci 29*: 331–364.

Sheetz M (2019) A tale of two states: Normal and transformed, with and without rigidity sensing. *Annu Rev Cell Dev Biol 35*: 169–190.

Shibutani ST, Saitoh, T, Nowag H, Münz C, Yoshimori T (2015) Autophagy and autophagy-related proteins in the immune system. *Nat Immunol 16*: 1014–1024.

Shipman P (2014) How do you kill 86 mammoths? Taphonomic investigations of mammoth megasites. *Quaternary International 359–360*: 38–46.

Shipton C, Blinkhorn J, Breeze PS, Cuthbertson P, Drake N, Groucutt HS, Jennings RP, Parton A, Scerri, EML Alsharekh A, Petraglia MD (2018) Acheulean technology and landscape use at Dawadmi, central Arabia. *PLOS ONE 13*(7): e0200497.

Shmerling RH (2018) Where people die. *Harvard Health Blog*. https://www.health.harvard.edu/blog/where-people-die-2018103115278.

Shultz M (1999) Microscopic investigation in fossil hominoidea: A clue to taxonomy, functional anatomy, and the history of diseases. *The Anatomical Record 257*: 225–232.

Siroky MB (2004) The aging bladder. *Rev Urol 6*(suppl 1): S3–S7.

Sjödin P, Sjöstrand AE, Jakobsson M, Blum MGB (2012) Resequencing data provide no evidence for a human bottleneck in Africa during the penultimate glacial period. *Molecular Biology and Evolution 29*: 1851–1860.

Smith E, Jacobs Z, Johnsen R, Ren M, Fisher EC, Oestmo S, Wilkins J, Harris JA, Karkanas P, Fitch S, Ciravolo A, Keenan D, Cleghorn N, Lane CS, Matthews T, Marean CW (2018) Humans thrived in South Africa through the Toba eruption about 74,000 years ago. *Nature 555*: 511–515.

Snyder-Beattie AE, Ord T, Bonsall MB (2019) An upper bound for the background rate of human extinction. *Sci Rep 9*: 11054.

Solly M (2019) Washington becomes first state to allow 'human composting' as a burial method. *Smithsonian Magazine* (April 23, 2019) https://www.smithsonianmag.com/smart-news/washington-first-state-allow-burial-method-human-composting-180972020/. Retrieved January 18, 2021.

Son M, Yu J, Kim KH (2015) Five questions about mycoviruses. *PloS Pathog 11*(11): e1005172.

Song H, Wignall PB, Chen ZQ, Tong J, Bond DPG, Lai X, Zhao X, Jiang H, Yan C, Niu Z, Chen J, Yang H, Wang Y (2011) Recovery tempo and pattern of marine ecosystems after the end-Permian mass extinction. *Geology 39*: 739–742.

Song S, Johnson FB (2018) Epigenetic mechanisms impacting aging: A focus on histone levels and telomeres. *Genes 9*: 201; doi: 10.3390/genes9040201.

Spellberg B, Bartlett JG, Gilbert DN (2013) The future of antibiotics and resistance. *N Engl J Med 368*: 299–302.

Srivastava D (2015) *Stem cells: The hype and the hope.* TEDxMarin, https://www.youtube.com/watch?v=HR7fLn2GGvI. Retrieved on January 18, 2021.

Statista (2020) *Global value of the cosmetics market 2018–2025.* https://www.statista.com/statistics/585522/global-value-cosmetics-market/. Retrieved on January 18, 2021.

Staubwasser M, Dragusin V, Onac BP, Assonov S, Ersek V, Hoffmann DL, Veres D (2018) Impact of climate change on the transition of Neanderthals to modern humans in Europe. *Proc Natl Acad Sci USA 115*: 9116–9121.

Stevens JC, Cain WS, Demarque A, Ruthruff AM (1991) On the discrimination of missing ingredients: aging and salt flavour. *Appetite 16*: 129–140.

Stolz A, Ernst A, Dikic I (2014) Cargo recognition and trafficking in selective autophagy. *Nat Cell Biol 16*: 495–501.

Storz P (2006) Reactive oxygen species–mediated mitochondria-to-nucleus signaling: A key to aging and radical-caused diseases. *Sci STKE 2006*(332): re3.

Sun G, Montell DJ (2017) Q&A: cellular near death experiences—what is anastasis? *BMC Biol 15*: 92.

Sun Y, Joachimski MM, Wignall PB, Yan C, chen Y, Jiang H, Wang L, Lai X (2012) Lethally hot temperatures during the early Triassic greenhouse. *Science 338*: 366–370.

Suzanne M, Steller H (2013) Shaping organisms with apoptosis. *Cell Death Differ20*: 669–675.

Szostak J, Bartel D, Luisi P (2001) Synthesizing life. *Nature 409*: 387–390.

Tachibana C (2019) Beyond CRISPR: What's current and upcoming in genome editing. *Science.* https://www.sciencemag.org/features/2019/09/beyond-crispr-what-s-current-and-upcoming-genome-editing#. Retrieved on January 18, 2021.

Takahashi K, Tanabe K, Lhnuki M, Narita M, Ichisaka T, Tomoda K, Yamanaka S (2007) Induction of pluripotent stem cells from adult human fibroblasts by defined factors. *Cell 131*: 861–872.

Takahashi K, Yamanaka, S (2006) Induction of pluripotent stem cells from mouse embryonic and adult fibroblast cultures by defined factors. *Cell 126*: 663–676.

Tang HM, Tang HL (2018) Anastasis: recovery from the brink of cell death. *R Soc Open Sci 5*: 180442.

Tang HM, Tang HL (2019) Cell recovery by reversal of ferroptosis. *Biol Open 8*(6): bio043182.

Tatar M, Bartke A, Antebi A (2003) The endocrine regulation of aging by insulin-like signals. *Science 299*: 1346–1351.

Tebb W, Vollum EP (1896) *Premature Burial and How It May Be Prevented.* Swan Sonnenschein and Co., London. http://www.gutenberg.org/files/50460/50460-h/50460-h.htm. Retrieved January 18, 2021.

Teno JM, Gozalo P, Trivedi AN, Bunker J, Lima J, Ogarek J, Mor V (2018) Site of death, place of care, and health care transitions among US Medicare beneficiaries, 2000–2015. *JAMA 320*: 264–271.

Tettamanti G, Casartelli M (2019) Cell death during complete metamorphosis. *Philosophical Transactions of the Royal Society, Part B 374*: 20190065.

Thomas H (2003) Do green plants age and if so how? *Top Curr Genet 3*: 145–171.

Thomas H (2013) Senescence, ageing and death of the whole plant. *New Phytologist 197*: 696–711.

Thomas HE, McKenzie MD, Angstetra E, Campbell PD, Kay TW (2009) Beta cell apoptosis in diabetes. *Apoptosis* 14: 1389–1404.

Thwaini A, Khan A, Malik A, Cherian J, Barua J, Shergill I, Mammen K (2006) Fournier's gangrene and its emergency management. *Postgrad Med J 82*: 516–519.

Tian X, Azpurua J, Hine C, Vaidya A, Myakishev-Rempel M, Ablaeva J, Mao Z, Nevo E, Gorbunova V, Seluanov A (2013) High-molecular-mass hyaluronan mediates the cancer resistance of the naked mole rat. *Nature 499*: 346–349.

Tilstra JS, Clauson CL, Niedernhofer LJ, Robbins PD (2011) NF-κB in aging and disease. *Aging Dis 2*: 449–465.

Tobin DJ (2017) Introduction to skin aging. *J Tissue Viability 26*: 37–46.

Tomczyk S, Fischer K, Austad S, Galliot B (2015) Hydra, a powerful model for aging studies. *Invertebr Reprod Dev 59* (sup1): 11–16.

Trautner JH, Reiser S, Blancke T, Unger K, Wysujack K (2017) Metamorphosis and transition between developmental stages in European eel (*Anguilla anguilla*, L.) involve epigenetic changes in DNA methylation patterns. *Comp Biochem Physiol Part D Genomics Proteomics 22*: 139–145.

Triffitt JT (2002) Stem cells and the philosopher's stone. *J Cell Biochem S38*: 13–19.

Tucci S, Vohr SH, McCoy RC, Vernot B, Robinson MR, Barbieri C, Nelson BJ, Fu W, Purnomo GA, Sudoyo H, Eichler EE, Barbujani G, Visscher PM, Akey JM, Green RE (2018) Evolutionary history and adaptation of a human pygmy population of Flores Island, Indonesia. *Science 361*: 511–516.

Turner L, Knoepfler P (2016) Selling stem cells in the USA: Assessing the direct-to-consumer industry. *Cell Stem Cell 19*: P154–P157.

Turner S, Gallois P, Brown D (2007) Trachery element differentiation. *Annu Rev Plant Biol 58*: 407–433.

UNESCO (1997) Universal Declaration on the Human Genome and Human Rights. United Nations Educational, Scientific, and Cultural Organization. Retrieved from: http://portal.unesco.org/en/ev.php-URL_ID=13177&URL_DO=DO_TOPIC&URL_SECTION=201.html. May 10, 2021.

UNICEF (2020) *Levels and trends in child mortality 2020*. UNICEF. https://www.unicef.org/reports/levels-and-trends-child-mortality-report-2020. Retrieved on January 18, 2021.

Uzelac B, Janošević D, Stojičić D, Budimir S (2017) Morphogenesis and developmental ultrastructure of *Nicotiana tabacum* short glandular trichomes. *Microscope Research Technique 80*: 779–786.

Vaesen K, Scherjon F, Hemerik L, Verpoorte A (2019) Inbreeding, Allee effects and stochasticity might be sufficient to account for Neanderthal extinction. *PLoS One 14*(11): e0225117.

van Deursen JM (2014) The role of senescent cells in ageing. *Nature 509*: 439–446.

Van Doorn WG, Beers EP, Dangl JL, Franklin-Tong VE, Gallois P, Hara-Nishimura I, Jones AM, Kawai-Yamada M, Lam E, Mundy J, Mur LAJ, Petersen M, Smertenko A, Tallansky M, Breusegem FV, Wolpert T, Woltering E, Zhivotovsky B, Bozhkov PV (2011) Morphological classification of plant cell deaths. *Cell Death Differ 18*: 1241–1246.

Van Etten JL (2011) Giant viruses. *American Scientist 99*: 304–311.

van Heemst D, Beekman M, Mooijaart SP, Heijmans BT, Brandt BW, Zwaan BJ, Slagboom PE, Westendorp RGJ (2005) Reduced insulin/IGF-1 signalling and human longevity. *Aging Cell 4*: 79–85.

van Leeuwen E, Cronin K, Haun D (2017) Tool use for corpse cleaning in chimpanzees. *Sci Rep 7*: 44091.

Van Lommel P, van Wees R, Meyers V, Elfferich I (2001) Near-death experience in survivors of cardiac arrest: a prospective study in the Netherlands. *Lancet 358*: 2039–2045.

van Regenmorte MHV (2016) The metaphor that viruses are living is alive and well, but it is no more than a metaphor. *Stud Hist Philos Sci C Biol Biomed Sci 59*: 117–124.

Vaskivuo TE, Tapanainen JS (2002) Apoptosis in the human ovary. *Reprod Biomed Online* 6: 24–35.

Vass AA (2001) Beyond the grave—understanding human decomposition. *Microbiol Today 28*: 190–192.

Vaupel JW (2003) The future of human longevity--How important are markets and innovation? *Sci. SAGE KE* 2003, pe18.

Veneault-Fourrey C, Barooah M, Egan M, Wakley G, Talbot NJ (2006) Autophagic fungal cell death is necessary for infection by the rice blast fungus. *Science 312*: 580–583.

Veneault-Fourrey C, Talbot NJ (2007) Autophagic cell death and its importance for fungal developmental biology and pathogenesis. *Autophagy 3*: 126–127.

Vespa J, Medina L, Armstrong DM (2020) *Demographic Turning Points for the United States: Population Projections for 2020 to 2060 Population Estimates and Projections Current Population Reports.* United States Census Bureau. P25–1144.

Vijg J, Le Bourg E (2017) Aging and the inevitable limit to human life span. *Gerontology 63*: 432–434.

Vila M, Przedborski S (2003) Targeting programmed cell death in neurodegenerative diseases. *Nat Rev Neurosci 4*: 365–375.

Villarreal LP, Witzany G (2010) Viruses are essential agents within the roots and stem of the tree of life. *Journal of Theoretical Biology 262*: 698–710.

Vincent JA (2006) Ageing contested: Anti-ageing science and the cultural construction of old age. *Sociol 40*: 681–698.

Vreeland RH, Rosenzweig WD, Powers DW (2000) Isolation of a 250 million-year-old halotolerant bacterium from a primary salt crystal. *Nature 407*: 897–900.

Vrselja Z (2019) Destined for destruction? *Science 366*: 46–49.

Vrselja Z, Daniele SG, Silbereis J, Talpo F, Morozov YM, Sousa AMM, Tanaka BS, Skarica M, Pletikos M, Kaur N, Zhuang ZW, Liu Z, Alkawadri R, Sinusas AJ, Latham SR, Waxman SG, Sestan N (2019) Restoration of brain circulation and cellular functions hours post-mortem. *Nature 568*: 336–343.

Walton KL, Johnson KE, Harrison CA (2017) Targeting TGF-β mediated SMAD signaling for the prevention of fibrosis. *Front Pharmacol 8*: 461. https://doi.org/10.3389/fphar.2017.00461.

Wanderley JLM, Thorpe PE, Barcinsn MA, Soong L (2013) Phosphatidylserine exposure on the surface of *Leishmania amazonensis* amastiogotes modulates in vivo infection and dendritic cell function. *Parasite Immunol 35*: 109–119.

Wang HY, Eguchi K, Yamashita T, Takahashi T (2020) Frequency-dependent block of excitatory neurotransmission by isoflurane via dual presynaptic mechanisms. *J Neurosci* DOI: https://doi.org/10.1523/JNEUROSCI.2946-19.2020.

Wang L, Cui J, Jin B, Zhao J, Xu H, Lu Z, Li W, Li X, Li L, Liang E, Rao X, Wang S, Fu C, Cao F, Dixon RA, Lin J (2020) Multifeature analyses of vascular cambial cells reveal longevity mechanisms in old *Ginkgo biloba* trees. *Proc Natl Acad Sci USA* doi/10.1073/pnas.1916548117.

Wardlaw T, Salama P, Brocklehurst C, Chopra M, Mason E (2010) Diarrhoea: Why children are still dying and what can be done. *The Lancet 375*: P870–872.

Wareham C (2015) Slowed ageing, welfare, and population problems. *Theor Med Bioeth 36*: 321–340.

Wargo MJ, Hogan DA (2006) Fungal—bacterial interactions: a mixed bag of mingling microbes. *Curr Opin Microbiol 9*: 359–364.

Watson CFI, Matsuzawa T (2018) Behaviour of nonhuman primate mothers toward their dead infants: uncovering mechanisms. *Philos Trans R Soc Lond B Biol Sci 373*: 20170261.

Watve M, Parab S, Jogdand P, Keni S (2006) Aging may be a conditional strategic choice and not an inevitable outcome for bacteria. *Proc Natl Acad Sci USA 103*: 14831–14835.

Weichhart T (2018) mTOR as regulator of lifespan, aging, and cellular senescence: A mini-review. *Gerontology 64*: 127–134.

Weyrich LS, Duchene S, Soubrier J, Arriola L, Llamas B, Breen J, Morris AG, Alt KW, Caramellis D, Dresely V, Farrell M, Farrer AG, Francken M, Gully N, Haak W, Hardy K, Harvati K, Held P, Holmes EC, Kaidonis J, Lalueza-Fox C, de la Fasilla M, Rosas A, Semal P, Soltysiak A, Townsend G, Usai D, Wahl J, Huson DH, Dobney K, Cooper A (2017) Neanderthal behaviour, diet, and disease inferred from ancient DNA in dental calculus. *Nature 544*: 357–361.

Wheeler T, von Braun J (2013) Climate change impacts on global food security. *Science 341*: 508–513.

Whinnery JE (1997) Psychophysiologic correlates of unconsciousness and near-death experiences. *Journal of Near-Death Studies 15*: 231–258.

White CJ, Marin M, Fessler DMT (2017) Not just dead meat: An evolutionary account of corpse treatment in mortuary rituals. *J Cognit Cult 17*: 146–168.

Whitman WB, Coleman DC, Wieve WJ (1998) Prokaryotes: The unseen majority. *Proc Natl Acad Sci USA 95*: 6578–6583.

WHO (1946) *Preamble to the Constitution of the World Health Organization*. New York, NY: WHO.

WHO (2012) *International guidelines for the determination of death—phase I*. World Health Organization. https://www.who.int/patientsafety/montreal-forum-report.pdf.

WHO (2019) Causes of death among children. World Health Organization. https://www.who.int/maternal_child_adolescent/data/causes-death-children/en/.

WHO (2020) The top 10 cuses of death. World Health Organization. https://www.who.int/news-room/fact-sheets/detail/the-top-10-causes-of-death.

Wijdicks EFM (2002) Brain death worldwide. *Neurology 58*: 20–25.

Wijdicks EFM (2010) The case against confirmatory tests for determining brain death in adults. *Neurology 75*: 77–83.

Williams GC (1957) Pleiotropy, natural selection, and the evolution of senescence. *Evolution 11*: 398–411.

Williams GN, Higgins MJ, Lewek MD (2002) Aging skeletal muscle: Physiologic changes and the effects of training. *Physical Therapy 82*: 62–68.

Wilson ML, Wrangham RW (2003) Intergroup relations in chimpanzees. *Annu Rev Anthropol 32*: 363–392.

Winegard T (2019) *The Mosquito: A Human History of Our Deadliest Predator*. Dutton, New York.

Witkamp FE, Zuylen L, Borsboom G, van der Rijt CCD, van der Heide A (2015) Dying in the hospital: What happens and what matters, according to bereaved relatives. *Journal of Pain and Symptom Management 49*: 203–213.

Witte DR, Grobbee DE, Bots ML, Hoes AW (2005) A meta-analysis of excess cardiac mortality on Monday. *European Journal of Epidemiology* (2005) *20*: 401–406.

Wittgenstein L (1922) *Tractatus Logico-Philosophicus*. Project Gutenberg, ebook no. 5740. http://www.gutenberg.org/ebooks/5740.

Wittig K, Kasper J, Seipp S, Leitz T (2011) Evidence for an instructive role of apoptosis during the metamorphosis of *Hydractinia echinata* (Hydrozoa). *Zoology 114*: 11–22.

Wittmann M, Neumaier L, Evrard R, Weibel A, Schmied-Knittel I (2017) Subjective time distortion during near-death experiences: An analysis of reports. *Zeitschrift für Anomalistik 17*: 309–320.

World Economic Forum (2019) *The Global Risks Report*, 14th Edition. https://www.weforum.org/reports/the-global-risks-report-2019. Retrieved on January 19, 2021.

Xin XF, Nomura K, Aung K, Welasquez AC, Yao J, Boutrot F, Chang JH, Zipfel C, He SY (2016) Bacteria establish an aqueous living space in plants crucial for virulence. *Nature 539*: 524–529.

Yamaguchi Y, Inouye M (2016) Toxin-antitoxin systems in bacteria and archaea. In *Stress and Environmental Regulation of Gene Expression and Adaptation in Bacteria, I&II* (Ed. FJ de Bruijn) chapter 2.7. Willey, New York. https://doi.org/10.1002/9781119004813.ch8. Retrieved on January 18, 2021.

Yamaguchi Y, Park JH, Inouye M (2011) Toxin-antitoxin systems in bacteria and archaea. *Annual Review of Genetics 45*: 61–79.

Yang B, Anderson JR, Li BG (2016) Tending a dying adult in a wild multi-level primate society. *Curr Biol 26*: R403–R404.

Yang J, McCormick MA, Zheng J, Xie Z, Tsuchiya M, Tsuchiyama S, El-Samad H, Ouyang Q, Kaeberlein M, Kennedy BK, Li H (2015) Systematic analysis of asymmetric partitioning of yeast proteome between mother and daughter cells reveals "aging factors" and mechanism of lifespan asymmetry. *Proc Natl Acad Sci USA 112*: 11977–11982.

Yang Z, Chen F, Alvarado JB, Benner SA (2011) Amplification, mutation, and sequencing of a six-letter synthetic genetic system. *J Am Chem Soc 133*: 15105–15112.

Yiping T, Sánchez-Iglesias S, Araújo-Vilar D, Fong LG, Young SG (2016) LMNA missense mutations causing familial partial lipodystrophy do not lead to an accumulation of prelamin A. *Nucleus 7*: 5: 512–521.

Yost CL, Jackson LJ, Stone JR, Cohn AS (2018) Subdecadal phytolith and charcoal records from Lake Malawi, East Africa imply minimal effects on human evolution from the ~74 ka Toba supereruption. *Journal of Human Evolution 116*: 75–94.

Younger S, Hyun I (2019) Pig brain study could fuel debates around death. *Nature 568*: 302–304.

Zahavi A (1975) Mate selection: a selection for a handicap. *J Theor Biol 53*: 205–214.

Zahn JM, Poosala S, Owen AB, Ingram DK, Lustig A, Carter A, Weeraratna AT, Taub DD, Gorospe M, Mazan-Mamczarz K, Lakatta EG, Boheler KR, Xu X, Mattson MP, Falco G, Ko MSH, Schlessinger D, Firman J, Kummerfeld SK, Wood WH 3rd, Zonderman AB, Kim SK, Becke KG (2007) AGEMAP: A gene expression database for aging in mice. *PLoS Genet 3*(11): e201.

Zeng TC, Aw AJ, Feldman MW (2018) Cultural hitchhiking and competition between patrilineal kin groups explain the post-Neolithic Y-chromosome bottleneck. *Nat Commun 9*: 2077.

Zhang P, Zhai Y, Cregg J, Ang KK-H, Arkin M, Kenyon C (2020) Stress resistance screen in a human primary cell line identifies small molecules that affect aging pathways and extend Caenorhabditis elegans' lifespan. *G3: Genes, Genomes, Genetics 20*: 849–862.

Zhang XH, Tee LY, Wang XG, Huang QS, Yang SH (2015) Off target effects in CRISPR/Cas0-mediated genome engineering. *Mol Ther Nucleic Acids 4*: e264.

Zhang Y, Lamb BM, Feldman AW, Zhou AX, Lavergne T, Li L, Romesberg FE (2017) A semisynthetic organism engineered for the stable expansion of the genetic alphabet. *Proc Natl Acad Sci USA 114*: 1317–1322.

Zhavoronkov A, Bhullar B (2015) Classifying aging as a disease in the context of ICD-11. *Front. Genet.*, https://doi.org/10.3389/fgene.2015.00326. Retrieved on January 19, 2021.

Zheng Y, Wang Y, Ding B, Fei Z (2017) Comprehensive transcriptome analyses reveal that potato spindle tuber viroid triggers genome-wide changes in alternative splicing, inducible *trans*-acting activity of phased secondary small interfering RNAs, and immune responses. *J Virol 91*:e00247–17.

Zhivotovsky LA, Rosenberg NA, Feldman MW (2003) Features of evolution and expansion of modern humans, inferred from genomewide microsatellite markers. *Am J Hum Genet 72*: 1171–1186.

Zhou Y, Zhang W, Liu Z, Wang J, Yuan S (2015) Purification, characterization and synergism in autolysis of a group of 1,3-b-glucan hydrolases from the pilei of *Coprinopsis cinerea* fruiting bodies. *Microbiol 161*: 1978–1989.

Zhou Z, He H, Wang K, Shi X, Wang Y, Su Y, Wang Y, Li D, Liu W, Zhang Y, Shen L, Han W, Shen L, Ding J, Shao F (2020) Granzyme A from cytotoxic lymphocytes cleaves GSDMB to trigger pyroptosis in target cells. *Science 368*: eaaz7548.

Zhua Q, Binghamb GP (2011) Human readiness to throw: the size–weight illusion is not an illusion when picking the best objects to throw. *Evolution and Human Behavior 32*: 288–293.

Zollikofer CPE, Ponce de León MS, Vandermeersch B, Lévêque F (2002) Evidence for interpersonal violence in the St. Césaire Neanderthal. *Proc Natl Acad Sci USA 99*: 6444–6448.

Zomorodipour A, Andersson SGE (1999) Obligate intracellular parasites: Rickettsia prowazekii and Chlamydia trachomatis. *FEBS Lett 452*: 11–15.

Zwicker D, Seyboldt R, Weber CA, Hyman AA, Julicher F (2017) Growth and division of active droplets: A model for protocells. *Nat Phys 13*: 408–413.

Index

For the benefit of digital users, indexed terms that span two pages (e.g., 52–53) may, on occasion, appear on only one of those pages.

Tables, figures and boxes are indicated by *t*, *f* and *b* following the page number